W0079814

Fluid Flow Measurement

Fluid Flow Measurement

Editor

Arti Kashyap

Fluid Flow Measurement
Edited by **Arti Kashyap**

Printed in 2017

ISBN: 978-1-68117-374-0

Library of Congress Control Number: 2015941562

© 2016 by
SCITUS Academics LLC,
616, Corporate Way, Suite 2, 4766,
Valley Cottage, NY 10989

www.scitusacademics.com

This book contains information obtained from highly regarded resources. Copyright for individual articles remains with the authors as indicated. All chapters are distributed under the terms of the Creative Commons Attribution License, which permits unrestricted use, distribution, and reproduction in any medium, provided the original author and source are credited.

Notice

Reasonable efforts have been made to publish reliable data and views articulated in the chapters are those of the individual contributors, and not necessarily those of the editors or publishers. Editors or publishers are not responsible for the accuracy of the information in the published chapters or consequences of their use. The publisher believes no responsibility for any damage or grievance to the persons or property arising out of the use of any materials, instructions, methods or thoughts in the book. The editors and the publisher have attempted to trace the copyright holders of all material reproduced in this publication and apologize to copyright holders if permission has not been obtained. If any copyright holder has not been acknowledged, please write to us so we may rectify.

Contents

Preface

The basic approach of the given measurement technique depends on the flowing medium (liquid/gas), nature of the flow (laminar/turbulent) and steady/unsteadiness of the medium. Accordingly, the fluid flow diagnostics are classified as measurement of local properties (velocity, pressure, temperature, density, viscosity, turbulent intensity etc.), integrated properties (mass and volume flow rate) and global properties (flow visualization). Also, these properties can be measured directly using certain devices or can be inferred from few basic measurements. For instance, if one wishes to measure the flow rate, then a direct measurement of volume/mass flow can be done during a fixed time interval. However, the secondary approach is to measure some other quantity such as pressure difference and/or fluid velocity at a point in the flow and then calculate the flow rate using suitable expressions. In addition, flow-visualization techniques are sometimes employed to obtain an image of the overall flow field. The parameters of interest for incompressible flow are the fluid viscosity, pressure/temperature, fluid velocity and its flow rate.

Editor

Chapter 1

Model Validation for Low and High Superficial Gas Velocity Bubble Column Flows

Deify Law[a], Francine Battaglia[a] and Theodore J. Heindel[b]

[a]Department of Mechanical Engineering, Virginia Polytechnic Institute and State University, Blacksburg, VA 24061, USA

[b]Department of Mechanical Engineering, Iowa State University, Ames, IA 50011, USA

ABSTRACT

In the present work, gas–liquid flow dynamics in a bubble column are simulated with CFDLib using an Eulerian–Eulerian ensemble-averaging method in a two-dimensional Cartesian system. The two-phase flow simulations are compared to experimental measurements of a rectangular bubble column performed by Mudde et al. [1997. Role of coherent structures on Reynolds stresses in a 2-D bubble column. A.I.Ch.E. Journal 43, 913–926] and a cylindrical bubble column performed by Rampure et al. [2003. Modeling of gas–liquid/gas–liquid–solid flows in bubble columns: experiments and CFD simulations. The Canadian Journal of Chemical Engineering 81, 692–706] for low and high superficial gas velocities, respectively. The objectives are to obtain grid-independent numerical solutions using CFDLib to reconcile unphysical results observed using FLUENT with increasing grid resolutions [Law, D., Battaglia, F., Heindel, T.J., 2006. Numerical simulations of gas–liquid flow dynamics in bubble columns. In: Proceedings of the ASME Fluids Engineering Division, IMECE2006-13544, Chicago, IL], and to validate computational fluid dynamics (CFD) simulations with experimental data to demonstrate the use of numerical simulations as a viable design tool for gas–liquid bubble column flows. Numerical predictions are presented for the local time-averaged liquid velocity and gas fraction at various axial heights as a function of horizontal or radial position. The effects of grid resolution, bubble pressure (BP) model, and drag coefficient models on the numerical predictions are examined. The BP model is hypothesized to account for bubble stability, thus providing physical solutions.

INTRODUCTION

Bubble column reactors are widely used in the chemical industry due to their excellent heat and mass transfer characteristics, simple construction, and ease of operation. These reactors are used in a variety of chemical processes such as Fischer-Tropsch synthesis, manufacture of fine chemicals, oxidation and alkylation reactions, effluent treatment, coal liquefaction, and fermentation reactions (Sanyal et al., 1999). Bubble column hydrodynamics are studied experimentally and computationally for scale-up and design considerations. The performance of bubble column reactors depends on the gas holdup (volumetric gas fraction), bubble size, bubble rise velocity, bubble–bubble interactions, and mixing rate (Joshi, 2001 and Joshi et al., 2002). Full-scale experimentation in bubble columns is expensive and therefore a more cost-effective approach for exploring these reactors is by using validated computational fluid dynamics (CFD) models.

Numerical simulations of bubble columns either employing Eulerian–Eulerian models (Monahan et al., 2005, Pan and Dudukovi , 2000, Rampure et al., 2003, Sanyal et al., 1999 and Sokolichin and Eigenberger, 1994), Eulerian–Lagrangian models (Delnoij et al., 1997a, Delnoij et al., 1997b and Delnoij et al., 1997c), or volume of fluid (VOF) (Lin et al., 1996) methods were surveyed. The Eulerian–Eulerian model treats dispersed (gas bubbles) and continuous (liquid) phases as interpenetrating continua, and describes the motion for gas and liquid phases in an Eulerian frame of reference. Sokolichin and Eigenberger (1994), Pan and Dudukovi (2000) and Monahan et al. (2005) performed two-dimensional (2D) simulations of gas–liquid flows for a rectangular bubble column using an Eulerian–Eulerian approach.Sokolichin and Eigenberger (1994) and Pan and Duduković (2000) predicted time-averaged axial liquid velocity and reported that the liquid primarily traveled up the column center and reduced gas holdup with liquid downflow at the walls. Monahan et al. (2005) demonstrated that the two-fluid model could predict flow transitions in bubble columns successfully. Other investigators such as Rampure et al. (2003), who conducted three-dimensional (3D) simulations, and Sanyal et al. (1999), who conducted axisymmetric simulations of a cylindrical bubble column reactor, used the Eulerian–Eulerian method as well. It should be noted that both Rampure et al. (2003) and Sanyal et al. (1999) conducted experiments in addition to numerical simulations and the time-averaged gas holdup simulations compared qualitatively with the experiments.

In the Eulerian–Lagrangian model, the continuous phase is described in an Eulerian representation while the dispersed phase is treated as discrete bubbles and each bubble is tracked by solving the equations of motion for individual bubbles. Delnoij et al.,

1997a, Delnoij et al., 1997b and Delnoij et al., 1997c applied an Eulerian–Lagrangian method to simulate detailed bubble–bubble interactions along with interfacial forces in the laminar bubbly-flow regime. They predicted a distinct bubble plume and a time-dependent, multiple-staggered vortex mode of circulation that characterized the liquid phase of the bubble column. The computed flow structure qualitatively compared with the experimentally observed flow patterns.

The VOF method solves the instantaneous Navier–Stokes equations to obtain the gas and liquid flow field with an extremely high spatial resolution. The evolution of the gas-liquid interface is tracked using a volume-tracking scheme. Lin et al. (1996) used the VOF method to provide time-dependent behavior of a dispersed bubbling flow and to account for the coupling effects of the pressure field and the liquid velocity on the bubble motion based on their 2D bubble column simulations. The computational results indicated the unsteady nature of the flow due to the coupling effects of the pressure field, liquid velocity, and bubble motion. The numerical results compared well both qualitatively and quantitatively with experiments (Lin et al., 1996).

The main advantage of the Eulerian–Lagrangian formulation comes from the fact that each individual bubble is modeled, allowing consideration of additional effects related to bubble–bubble and bubble–liquid interactions. Mass transfer with and without chemical reaction, bubble coalescence and redispersion, in principle, can be added directly to an Eulerian–Lagrangian hydrodynamic model. The Eulerian–Lagrangian approach, which requires tracking the dynamics of each bubble, is usually applied to cases where a small number of bubbles exist, such as when the superficial gas velocity is low, due to computer limitations. On the other hand, the Eulerian–Eulerian method is often used because memory storage requirements and demand of computer power depend on the number of computational cells considered instead of the number of bubbles. The Eulerian–Eulerian approach can be applied to cases for low and high superficial gas velocities. The disadvantage of using the Eulerian–Eulerian method is that the bubble–bubble and bubble–liquid interactions cannot be considered as straightforward as the Eulerian–Lagrangian method. The VOF method is the most detailed model used to advance the gas–liquid interface through the flow field in an Eulerian mesh and does not require empirical constitutive equations. However, the VOF method is limited to a small number of bubbles e.g., less than ~10 bubbles in the flow field due to computational limitations. Most industrial applications require high superficial gas velocities and therefore the Eulerian–Eulerian method is preferred (Pan and Dudukovi , 2000).

Law et al. (2006) computationally studied bubble column hydrodynamics using the Eulerian–Eulerian method formulated in FLUENT. Unphysical results were observed with increasing grid resolutions, therefore prohibiting a conclusive grid resolution study. An alternate multiphase FORTRAN code, CFDLib, developed at Los Alamos National Laboratory, is tested in the present work. The hypothesis is that a bubble pressure (BP) model will provide numerical stability and more accurately represent the bubble phase in order to resolve the flow field correctly. For example, Monahan et al. (2005) reported that the BP model provided a uniform time-averaged gas holdup profile across the bubble column in the low superficial gas velocity (homogeneous) flow regime. The influence of closure models for drag and virtual mass used within the framework of CFDLib was investigated by Rafique and Duduković (2006) for bubble column flows. Their findings showed that most correlations agreed well with experimental data found in the literature, but they were unable to recommend a specific model due to the lack of accurately measured gas holdup profiles.

In the present work, the gas–liquid flow dynamics in a bubble column are simulated using CFDLib in 2D Cartesian coordinates and compared to previous results using FLUENT. The simulations are also compared to the experimental measurements of a cylindrical bubble column with a high superficial gas velocity performed by Rampure et al. (2003), and a rectangular bubble column with a low superficial gas velocity performed by Mudde et al. (1997). Numerical predictions are presented for the time-averaged phase variables at various axial heights as a function of radial or horizontal position. The effects of grid resolution, BP model, and drag coefficient model on the numerical predictions are also examined. The specific objectives of this study are to obtain grid-independent numerical solutions and to validate the CFD simulations with the published experimental data in order to demonstrate the use of numerical simulations as a viable design tool.

NUMERICAL FORMULATION

Governing Equations

CFDLib, a FORTRAN code developed at Los Alamos National Laboratory, uses a finite-volume technique to integrate the time-dependent equations of motion that govern multiphase flows. The code is based on an arbitrary Lagrangian–Eulerian (ALE) scheme as described by Hirt et al. (1974). The name ALE refers to the flexibility of the scheme, which allows for the mesh either to be moved along with the fluid (Lagrangian), to remain in a fixed position (Eulerian), or to be moved in another fashion as selected by the user. The ALE scheme is designed to handle flows at any flow speed, including incompressible and hypersonic flows, and allows for multiphase calculations for an arbitrary number of fluid fields. The two-fluid Eulerian–Eulerian model is employed to represent each phase as interpenetrating continuum and the conservation equations for mass and momentum for each phase are ensemble-averaged. The subscript c refers to the continuous (liquid water) phase and the subscript d refers to the dispersed (air bubble) phase. The continuity equations for each phase, neglecting mass transfer, are

$$\frac{\partial}{\partial t}(\alpha_c \rho_c) + \nabla \cdot (\alpha_c \rho_c \vec{u}_c) = 0$$

(1)

$$\frac{\partial}{\partial t}(\alpha_d \rho_d) + \nabla \cdot (\alpha_d \rho_d \vec{u}_d) = 0$$

(2)

The momentum equations for each phase are

$$\frac{\partial}{\partial t}(\alpha_c \rho_c \vec{u}_c) + \nabla \cdot (\alpha_c \rho_c \vec{u}_c \vec{u}_c)$$

$$= -\alpha_c \nabla p + \nabla \cdot \bar{\bar{\tau}}_c + \vec{K}_{dc}(\vec{u}_d - \vec{u}_c) + \vec{F}_{vm} + \rho_c \alpha_c \vec{g}$$

(3)

$$\frac{\partial}{\partial t}(\alpha_d \rho_d \vec{u}_d) + \nabla \cdot (\alpha_d \rho_d \vec{u}_d \vec{u}_d)$$

$$= -\alpha_d \nabla p + \nabla \cdot \bar{\bar{\tau}}_d + \vec{K}_{cd}(\vec{u}_c - \vec{u}_d) - \vec{F}_{vm} + \rho_d \alpha_d \vec{g}$$

(4)

The terms on the right hand side of Eqs. (3) and (4) represent, from left to right, the pressure gradient, effective stress, interfacial momentum exchange (drag and virtual mass forces), and the gravitational force. The closures for turbulence modeling and interfacial momentum exchange are discussed next.

Turbulence Modeling

Turbulence contributions for the continuous and the dispersed phases are modeled through a set of modified standard k–ꞷ equations that calculate the turbulence generated at the gas–liquid interface in the form of a slip-production energy term (Kashiwa and VanderHeyden, 2000 and Launder and Spalding, 1974). The modified k–ꞷ equations are used only for the high superficial gas velocity case and the equations for a general phase k are

$$\frac{\partial}{\partial t}(\alpha_k \rho_k k_k) + \nabla \cdot (\alpha_k \rho_k k_k \vec{u}_k)$$

$$= \nabla \cdot \left(\alpha_k \frac{\mu_{t,k}}{\sigma_k} \nabla k_k \right) + \alpha_k G_k - \alpha_k \rho_k \varepsilon_k$$

$$+ \sum_{l \neq k} \beta_{kl} K_{kl} |\vec{u}_k - \vec{u}_l|^2 + 2 \sum_{l \neq k} E_{kl}(k_l - k_k)$$

(5)

$$\frac{\partial}{\partial t}(\alpha_k \rho_k \varepsilon_k) + \nabla \cdot (\alpha_k \rho_k \varepsilon_k \vec{u}_k)$$

$$= \nabla \cdot \left(\alpha_k \frac{\mu_{t,k}}{\sigma_\varepsilon} \nabla \varepsilon_k \right) + \alpha_k \frac{\varepsilon_k}{k_k}(C_{1\varepsilon}G_k - C_{2\varepsilon}\rho_k \varepsilon_k)$$

$$+ \frac{1}{\tau_{kl}}\left\{ \sum_{l \neq k} \beta_{kl} K_{kl} |\vec{u}_k - \vec{u}_l|^2 \right\} \tag{6}$$

Where

$$\mu_{t,k} = \rho_k C_{\mu,k} \frac{k_k^2}{\varepsilon_k} \tag{7}$$

$$G_k = \mu_{t,k}(\nabla \vec{u}_k + (\nabla \vec{u}_k)^T) : \nabla \vec{u}_k \tag{8}$$

$$C_{\mu,k} = \frac{C_\mu}{1 + (2\sum_{l \neq k} E_{kl} k_k)/(\rho_k \varepsilon_k)} \tag{9}$$

The form of Eq. (9) models a return-to-isotropy effect due to fluctuating interfacial momentum coupling and reduces the turbulent viscosity from that predicted by the single-phase model. The turbulence energy exchange rate coefficient E_{kl} is given by

$$E_{kl} = \alpha_k \alpha_l \left(\frac{\rho_k \rho_l}{\rho_k + \rho_l} \right) \frac{\sqrt{k_k + k_l}}{d_b}(1 + Re^{0.6}). \tag{10}$$

where, Re is the bubble Reynolds number and the expression will be discussed in Section 2.4.

The first three terms on the right-hand side of Eq. (5) account for turbulent diffusion, mean flow shear production, and decay of turbulence kinetic energy of phase k. The fourth term on the right-hand side of Eq(5) accounts for production of turbulence energy from slip between phases. The coefficient β_{kl} is given by

$$\beta_{kl} = \frac{a_k}{a_k + a_l}$$
(11)

Where

$$a_k = \frac{\alpha_k^{1/3}}{\rho_k + \rho_c}$$
(12)

and ρ_c is the continuous phase density. The last term in Eq. (5) accounts for the exchange of turbulence energy among phases.

The first three terms on the right-hand side of Eq. (6) account for the diffusion of turbulence dissipation, the mean flow velocity gradient production term, and the homogeneous dissipation term. The last group of terms in Eq. (6) describes the effect of interfacial momentum transfer on the production of turbulence dissipation. The time constant τ_{kl} is given by the following empirical correlation:

$$\tau_{kl} = \left\{ 0.01 C_{2\varepsilon} (\alpha_k \alpha_l)^{0.086} \left[\frac{\rho_k |\vec{u}_k - \vec{u}_l| d_b}{\mu_k} \right]^{0.562} \right.$$
$$\left. \times \frac{|\vec{u}_k - \vec{u}_l|}{d_b} \right\}^{-1}$$
(13)

This correlation was obtained by fitting predictions of turbulence kinetic energy to data from experiments on homogeneous sedimenting and bubbly systems (Lance and Bataille, 1991 and Mizukami et al., 1992; Parthasarathy and Faeth, 1990a and Parthasarathy and Faeth, 1990b). The term $_{Kkl}$ is the momentum exchange coefficient and the model will be discussed in Section 2.3. Eqs. (7) and (8) are closure models for the turbulent viscosity $_{\mu t,k}$ and the production of turbulent kinetic energy $_{Gk}$ of phase k . The turbulent parameters are set using standard empirical values for k–turbulence modeling where $C_1=1.44$, $C_2 =1.92$, $_{C\mu}=0.09$, $_{\sigma k}=1.0$, and $\sigma=1.3$.

Interfacial Momentum Exchange

The interfacial momentum exchange terms in the momentum conservation equations for each phase consist of drag and virtual mass force terms. The drag force for the gas and liquid is modeled, respectively, as

$$\vec{K}_{cd}(\vec{u}_c - \vec{u}_d) = \frac{3}{4}\rho_c \alpha_d \alpha_c \frac{C_D}{d_b}|\vec{u}_c - \vec{u}_d|(\vec{u}_c - \vec{u}_d)$$

$$\vec{K}_{dc}(\vec{u}_d - \vec{u}_c) = \frac{3}{4}\rho_c \alpha_d \alpha_c \frac{C_D}{d_b}|\vec{u}_d - \vec{u}_c|(\vec{u}_d - \vec{u}_c)$$

(14)

where C_D is the drag coefficient. The virtual mass force $\vec{F}_{vm}$ is modeled as

$$\vec{F}_{vm} = 0.5\alpha_d \rho_c \left(\frac{d\vec{u}_c}{dt} - \frac{d\vec{u}_d}{dt} \right)$$

(15)

and the coefficient of 0.5 is used for a spherical bubble.

Drag Coefficient Model

Two drag coefficient models are used in this study for gas–liquid flows. The drag coefficient model proposed by Schiller and Naumann (1933) is

$$C_D = \begin{cases} 24(1 + 0.15Re^{0.687})/Re, & Re \leqslant 1000 \\ 0.44, & Re > 1000 \end{cases}$$

(16)

where $Re = \rho_c |\vec{u}_d - \vec{u}_c| d_b / \mu_c$ is the bubble Reynolds number based on a characteristic (effective) bubble diameter, relative velocity between the two phases, and the liquid density and dynamic viscosity. Another drag coefficient relation proposed by White (1974) and used in this study is expressed by

$$C_D = C_\infty + \frac{24}{Re} + \frac{6}{1 + \sqrt{Re}}, \quad 0 \leqslant Re \leqslant 2 \times 10^5 \tag{17}$$

where C_∞ is the drag coefficient when the bubble Reynolds number goes to infinity. In our calculation C_∞ is set to 0.5.

Bubble Pressure Model

The BP model represents the transport of momentum arising from bubble-velocity fluctuations, collisions, and hydrodynamic interactions. The BP model is reported in the literature to play an important role in bubble-phase stability (Spelt and Sangani, 1998). Biesheuvel and Gorissen (1990) proposed a BP model of the form

$$P_d = \rho_c C_{BP} \alpha_d (u_d - u_c)(u_d - u_c) \left(\frac{\alpha_d}{\alpha_{dcp}} \right) \left(1 - \frac{\alpha_d}{\alpha_{dcp}} \right) \tag{18}$$

The gradient of Eq. (18) is added to the right-hand side of the gas momentum Eq. (4). A positive value of $dP_d/d\alpha_d$ acts as a driving force for bubbles to move from areas of higher α_d to areas of lower α_d and facilitates stabilization of the bubbly-flow regime. The virtual mass coefficient C_{BP} of an isolated spherical bubble is 0.5. The BP is proportional to the slip velocity and gas holdup. The gas holdup at close packing α_{dcp} is set equal to 1.0 in this study.

As we will show for the low superficial gas velocity case, the BP model must be employed with a bubble induced turbulence (BIT) model to obtain numerical stability at high grid resolutions. Sato et al. (1981) proposed a BIT model proportional to the bubble diameter and slip velocity of the rising bubbles

$$\mu_{t,c} = \rho_c C_{BT} \alpha_d d_b |\vec{u}_d - \vec{u}_c| \tag{19}$$

where the value of the proportionality constant C_{BT} is 0.6. Eq. (19) replaces Eq. (7) when the BIT model is applied. The BIT model yields an effective viscosity in the liquid (continuous) phase that is the sum of the molecular viscosity of the continuous phase and the turbulent viscosity calculated from the BIT model, whereas the effective viscosity for the dispersed phase is assumed to equal the molecular viscosity of the dispersed phase.

Simulation Conditions

The marker and cell (MAC) method has been selected in CFDLib to solve the incompressible gas–liquid two-phase flow. A velocity inlet boundary condition is used to introduce gas flow uniformly at the bottom of the bubble column. A no-slip boundary condition is applied for both phases at the walls and ambient pressure is specified at the top of the domain. The convergence criterion is set to 1×10^{-8} for all dependent variables. All simulations use a time step according to the Courant–Friedrich–Lewy (CFL) stability criterion to march the solution forward and the results are time-averaged from 20 to 90 s, which includes 7000 time realizations. To be consistent with the findings of Rampure et al. (2003) and Mudde et al. (1997), an effective bubble size of 0.5 cm is used to represent the dispersed gas phase.

RESULTS AND DISCUSSION

High Superficial Gas Velocity

Simulations are performed to match the experimental conditions of Rampure et al. (2003) for a cylindrical bubble column operating in the heterogeneous flow regime with H_d=1.0m and W=0.2m, as shown inFig. 1. The static water height $(_{Hc})$ in the column is 1.0 m and air flows uniformly through the bottom of the column at 0.1 m/s. The geometry in Fig. 1 is modeled in Cartesian coordinates as a 2D slice through the centerplane of the cylinder. The application of a 2D Cartesian simulation has been shown to provide adequate representation of the flow physics when compared to 3D cylindrical and rectangular bubble column simulations (Law et al., 2006 and Monahan et al., 2005).

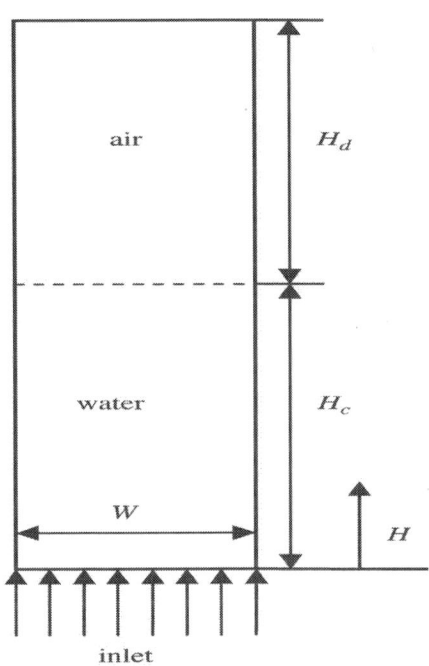

Figure 1: Schematic of the simulation domain used in this study.

Grid Resolution Study

Six different grid resolutions are tested using CFDLib to simulate the 2D representation of the domain inFig. 1. The number of computational cells along the horizontal and vertical directions is increased while maintaining a square computational cell; a summary of grid resolutions is presented in Table 1. The effects of grid resolution on the numerical predictions of the time-averaged gas holdup are examined and compared to previous results by the authors using FLUENT (Law et al., 2006). For the FLUENT and CFDLib simulations, the Schiller–Naumann drag coefficient model is used.

Table 1: Number of cells and cell size x- and y-directions for the high superficial gas velocity simulations

Number of Cells (x×y)	Δx=Δy (cm)
10×100	2.00
15×150	1.33
20×200	1.00
30×300	0.66
40×400	0.50
60×600	0.33

Fig. 2 shows the predictions of the time-averaged gas holdup using FLUENT for axial locations ofH=0.15 and 0.65 m above the distributor plate (Law et al., 2006). Results from the experiments ofRampure et al. (2003) are compared to the simulations for varying grid resolution, and the simulations are plotted in terms of r=x/2. FLUENT best predicts the experimental data using the 30 × 300 grid resolution whereas the 10×100 grid resolution gives completely unphysical flow dynamics that fail to capture experimental phenomena. For the cases of 15×150 and 20×200 cells, the numerically predicted values overestimate the maximum gas holdup at the centerline and underestimate it toward the walls. Grid-independent solutions for this gas–liquid bubble column flow were not obtained (Law et al., 2006). FLUENT was not able to satisfy convergence for the resolution of 60×600 cells, even with a small time step size of 0.0001 s. The radial profiles for all predicted time-averaged gas holdup values using FLUENT are generally symmetric, except for the lowest resolution case of 10×100 cells.

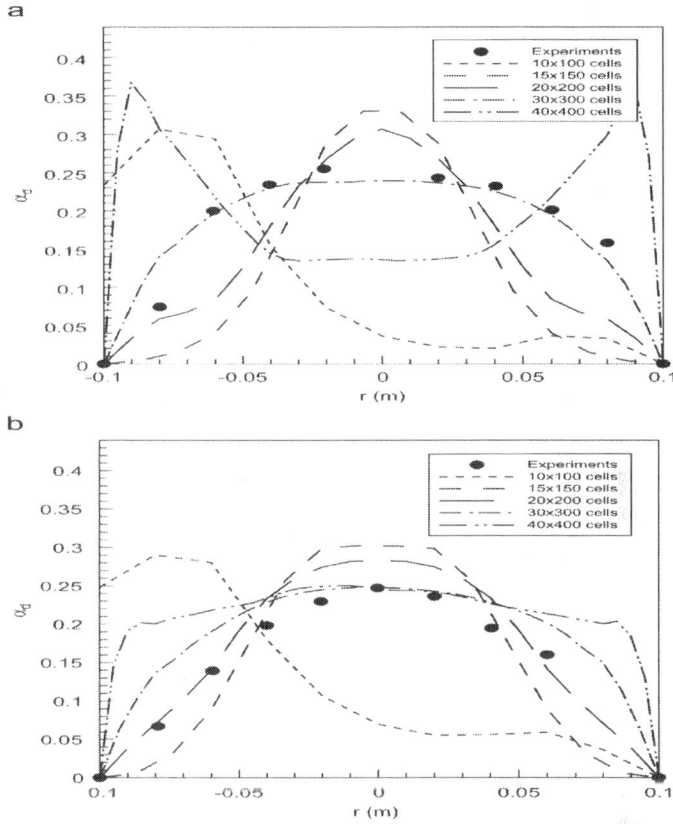

Figure 2: Time-averaged gas holdup versus radial position, comparing FLUENT simulations (Law et al., 2006) using different grid resolutions to experimental data by Rampure et al. (2003) at heights of (a) H=0.15m and (b) H=0.65m.

Similar trends are observed from the CFDLib simulations, shown in Fig. 3 for axial heights of H=0.15 and 0.65 m. At low axial heights, both FLUENT (Fig. 2a) and CFDLib (Fig. 3a) overpredict the gas holdup near the centerline for low grid resolutions, e.g., 15×150 grid cells. For higher grid resolutions such as40×400 cells, both codes predict two local maxima for the gas holdup. The gas holdup predictions using CFDLib at H=0.65m (Fig. 3b) indicate grid convergence for the four largest grid resolutions, which was not observed when using FLUENT. Fig. 4 presents time-averaged gas holdup contours for both FLUENT and CFDLib that elucidate the large gas-deficient region in the center of the column near the inlet.

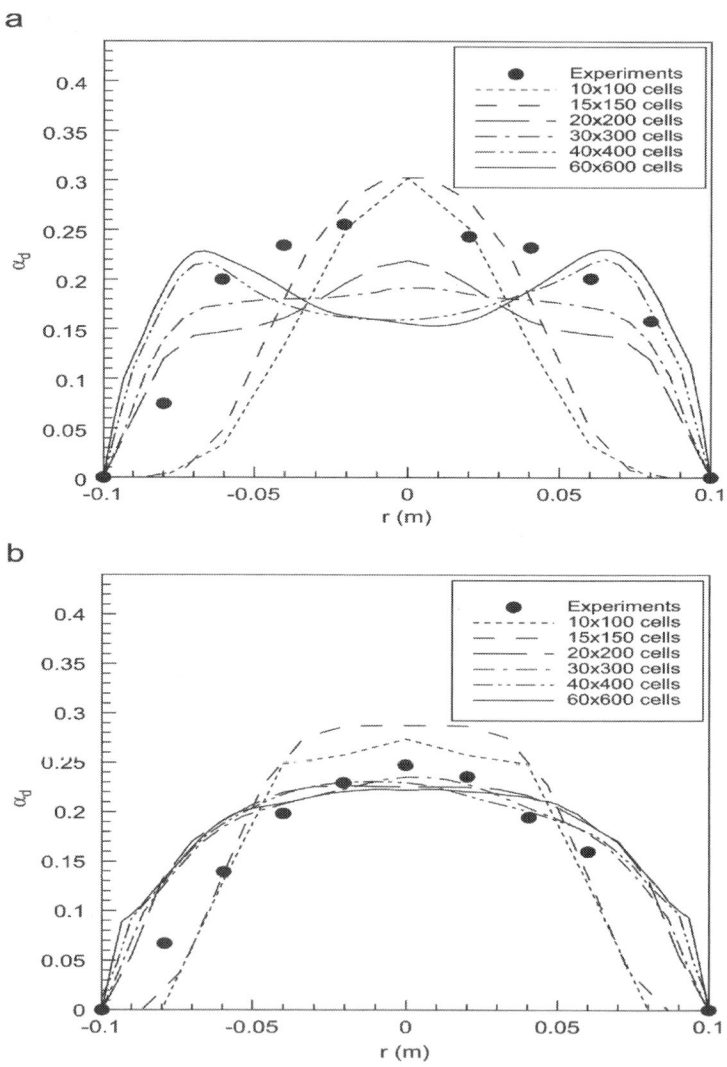

Figure 3: Time-averaged gas holdup versus radial position, comparing CFDLib simulations using different grid resolutions to experimental data by Rampure et al. (2003) at heights of (a) H=0.15m and (b) H=0.65m.

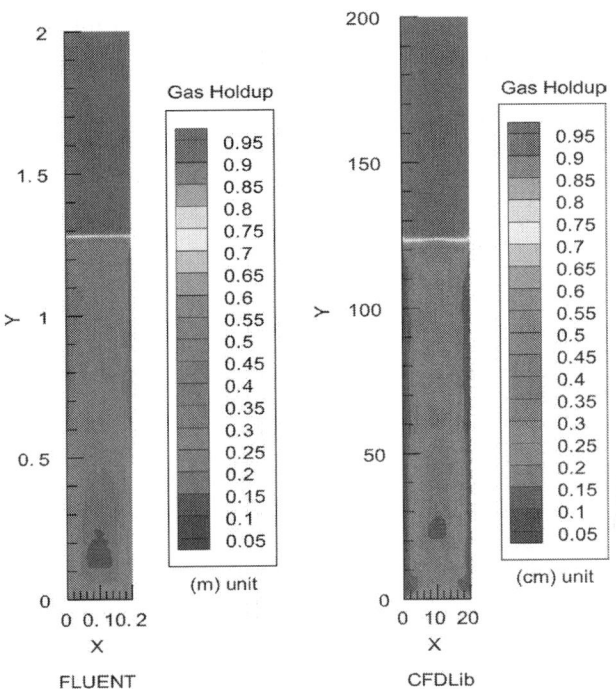

Figure 4: Time-averaged gas holdup contour plot, comparing FLUENT with CFDLib simulations for the high superficial gas velocity case using 40×400 cells.

Bubble Pressure Model Study

The effect of the BP model using CFDLib is examined to determine how it affects flow stability. The grid resolutions given in Table 1 are used and new simulations for the time-averaged gas holdup employing the BP model are presented here. Fig. 5a shows that as grid resolutions become finer, two off-center high gas holdup regions are observed, which is similar to the FLUENT simulations (Fig. 2a) and CFDLib simulations without the BP model (Fig. 3a) at the lower axial height. The influence of the BP model at the higher axial height (Fig. 5b) is also negligible. Fig. 6 directly compares simulations with and without the BP model for several grid resolutions. It is interesting to note that gas holdup predictions at the higher axial location are very similar with and without the BP model for the three largest grid resolutions. A grid resolution of80×800 was also tested using CFDLib but the simulations diverged, even with the BP model. The previous work of Law et al. (2006) suggested that cell sizes smaller than the effective bubble diameter may cause numerical stability issues and result in a diverged simulation. Furthermore, the flow conditions are indicative of a turbulent flow, thus it is not surprising that the BP model had little effect.

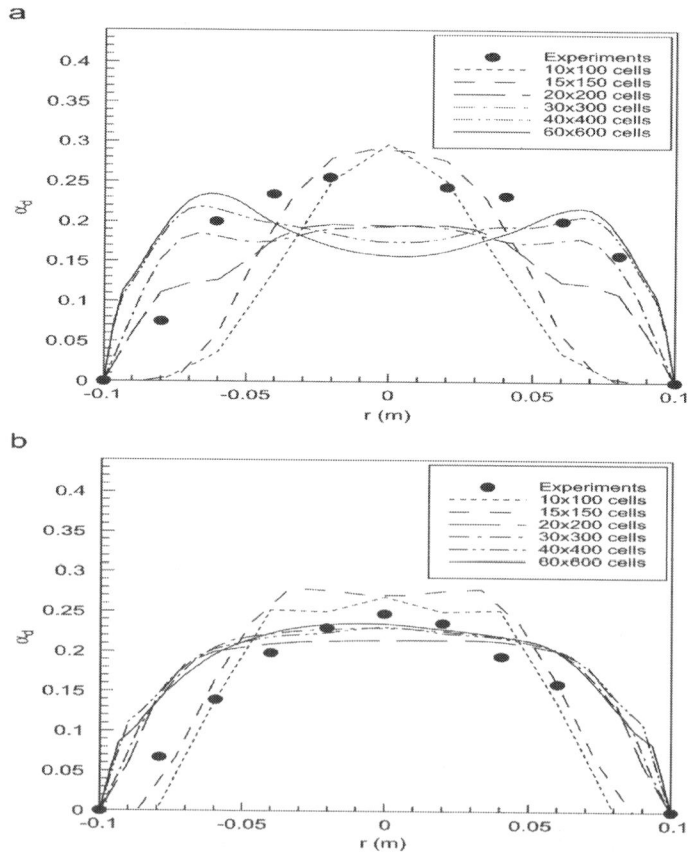

Figure 5: Time-averaged gas holdup versus radial position, comparing CFDLib simulations using the bubble pressure model to experimental data by Rampure et al. (2003) at heights of (a) H=0.15m and (b) H=0.65m.

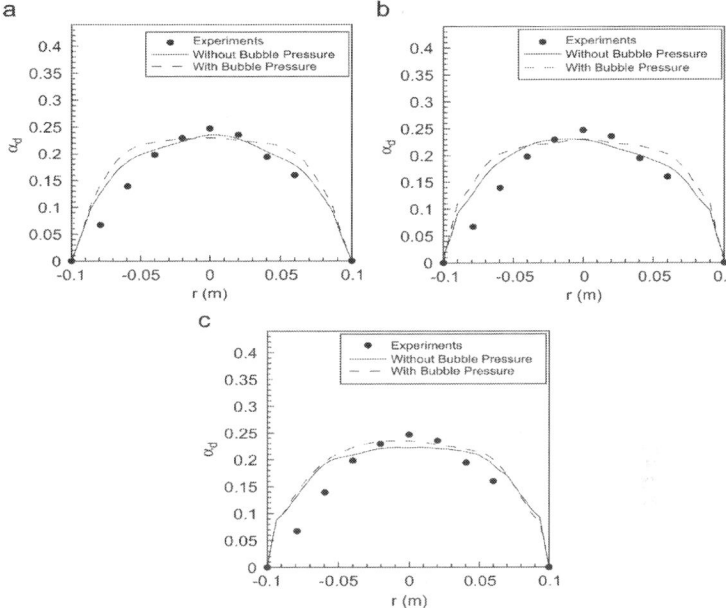

Figure 6: Time-averaged gas holdup versus radial position, comparing CFDLib simulations with and without the bubble pressure model to experimental data by Rampure et al. (2003) at a height of H=0.65m using (a) 30×300, (b) 40×400, and (c) 60×600 grid cells.

Drag Coefficient Model Study

Drag coefficient models are investigated to determine the effects on gas holdup. The grid resolution for this portion of the study is 40×400 grid cells and the BP model is not employed. Fig. 7 presents the predicted time-averaged gas holdup for the Schiller–Naumann and White models at two axial heights. The White model compares very well with the experimental results (Rampure et al., 2003) at the lower axial height. However, the Schiller–Naumann model compares better to the experiments than the White model at the higher axial liquid height, predicting a more parabolic profile (Fig. 7b). As an additional comparison, Fig. 8 presents predictions using CFDLib employing the White model, with and without the BP model. It is evident that the BP model has no discernible affects on the bubble dynamics. In general, the time-averaged gas holdup profiles are more homogeneous using the White model as compared to the Schiller–Naumann model.

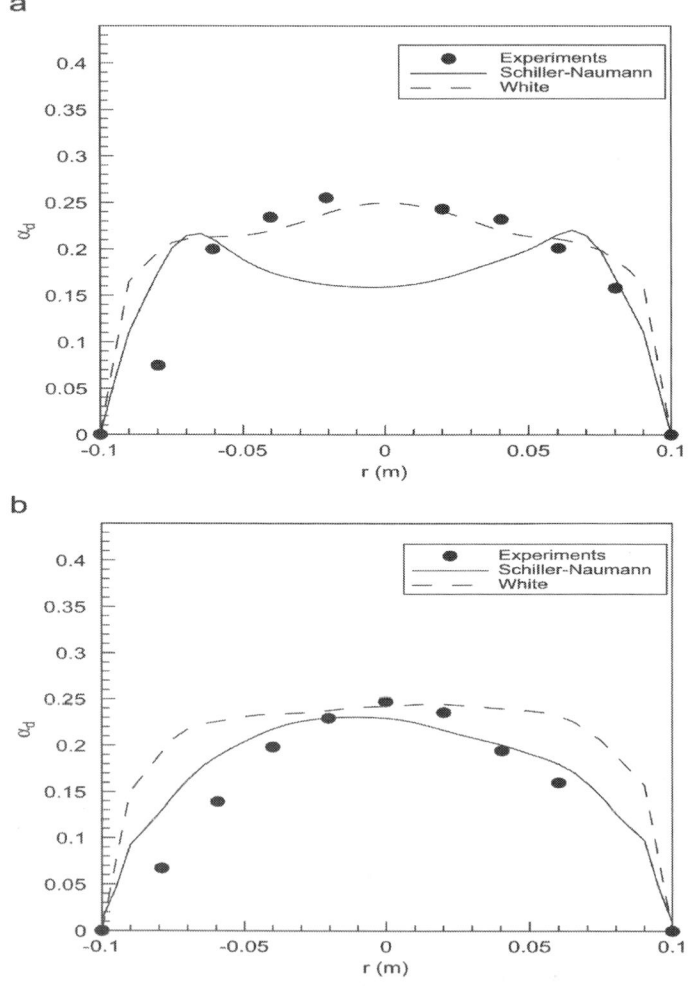

Figure 7: Time-averaged gas holdup versus radial position, comparing CFDLib simulations using the Schiller–Naumann and White drag coefficient models for 40×400 grid cells without using the BP model to experimental data by Rampure et al. (2003) at heights of (a) H=0.15m and (b) H=0.65m.

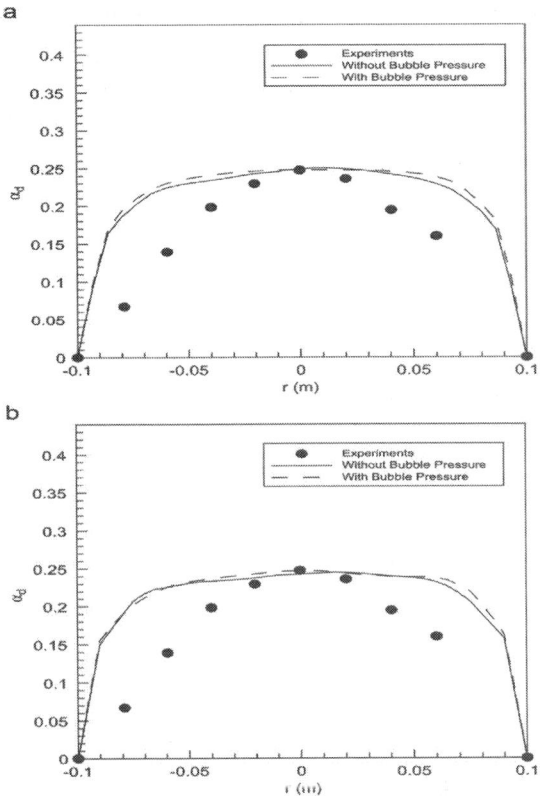

Figure 8: Time-averaged gas holdup versus radial position, comparing CFDLib simulations using the White drag coefficient model with and without the BP model to experimental data by Rampure et al. (2003) at a height of H=0.65m using (a) 30×300 and (b)40×400 grid cells.

Fig. 9 presents time-averaged gas holdup contours for a grid resolution of 40×400 cells without using the BP model, comparing the two drag models. Two high gas holdup regions form off-center using the Schiller–Naumann model at lower axial heights, whereas the White model predicts a more homogeneous gas holdup throughout the entire liquid bed. Furthermore, the bed expansion is higher with the White model than with the Schiller–Naumann model. The results suggest that the accuracy of the drag models may be dependent on the flow regime and further studies in the next section will explore this notion.

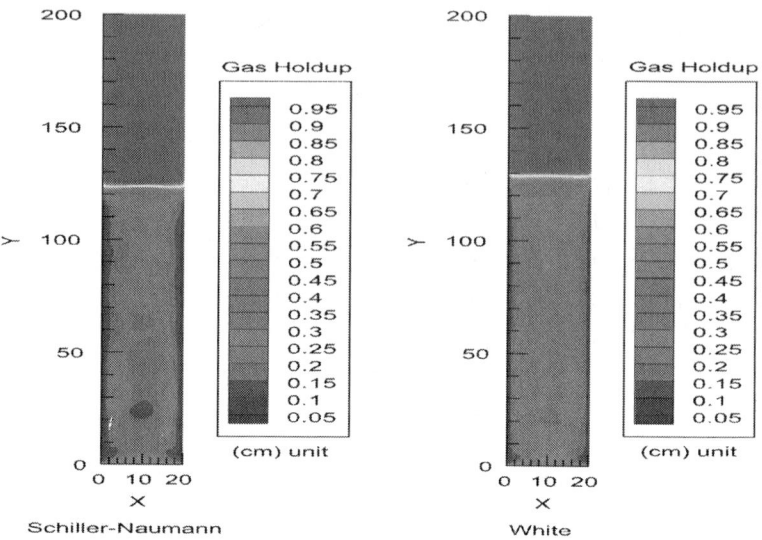

Figure 9: Time-averaged gas holdup contour plots comparing CFDLib simulations using Schiller–Naumann and White drag coefficient models for 40×400 grid cells without using the BP model.

Low Superficial Gas Velocity

Simulations are performed to match the experimental conditions of Mudde et al. (1997) for a rectangular bubble column that is H_c=1.1m, H_d=0.4m, and W=0.5m (see Fig. 1). The rectangular column used in the experiments of Mudde et al. (1997) had a depth of 0.0127 m; however, in the present study the column is modeled in a 2D Cartesian system. Air flows uniformly through the bottom of the column at 0.01 m/s, which is 10 times slower than the high superficial gas velocity case.

Grid Resolution Study

Three different grid resolutions are tested using FLUENT and CFDLib to simulate the 2D representation of the Mudde et al. (1997) domain in Fig. 1. The effects of grid resolution on the numerical predictions of the time-averaged gas holdup are examined by employing the Schiller–Naumann drag coefficient model without using the BP model. The number of computational cells along the horizontal and vertical directions is increased while maintaining a square computational cell. Table 2 depicts that for the coarse and medium grid resolutions, where the cell size is larger than the effective bubble diameter (0.5 cm), the simulations converge regardless of numerical code using the Schiller–Naumann drag coefficient model without the BP model. However, neither FLUENT nor CFDLib is able to predict a converged solution for 60×600 cells when using the Schiller–Naumann drag model without the BP model.

Table 2: Grid resolutions and models tested for the low superficial gas velocity case

Grid resolution	Cell size (cm)	FLUENT SN model	CFDLib SN model			CFDLib White model		
			Without BP	BP	BP and BIT	Without BP	BP	BP and BIT
15×150	1.00	C	C	C	C	C	C	C
30×300	0.50	C	C	C	C	C	C	C
60×600	0.25	D	D	D	C	D	D	C

Key: SN: Schiller–Naumann; BP: Bubble pressure model; BIT: Bubble induced turbulence model; C: Converged solution; D: Diverged solution.

Bubble Pressure Model Study

Table 2 depicts that when the BP model is implemented in CFDLib using the Schiller–Naumann drag coefficient model, the 60×600 cell solution still diverges. Table 2 further depicts that the only means by which a stable time-dependent solution is possible with the 60×600 grid cell resolution is by using CFDLib, while employing both the BP and BIT models, as recommended by Monahan et al. (2005) andMonahan and Fox (2007). This study confirms that the BP and BIT models provide numerical stability to the numerical predictions of gas–liquid bubble columns under low superficial gas velocity flows when the flow is characterized as homogeneous. Hence, the remaining figures show simulations with both the BP and BIT models employed in CFDLib. Fig. 10 compares experimental results of Mudde et al. (1997) with the predictions of the time-averaged axial liquid velocity using CFDLib with the Schiller–Naumann drag coefficient model for axial locations of H=0.30 and 0.75 m above the distributor plate. Simulations for the30×300 and 60×600 grids are shown. The simulations for both grids compare well with the experiments at H=0.30m (Fig. 10a), but more variation between the simulations and experiments is observed at the center of column when H=0.75m (Fig. 10b). The discrepancy is due to experimental error as reported by Mudde et al. (1997) and was also observed in the work by Rafique and Duduković (2006).

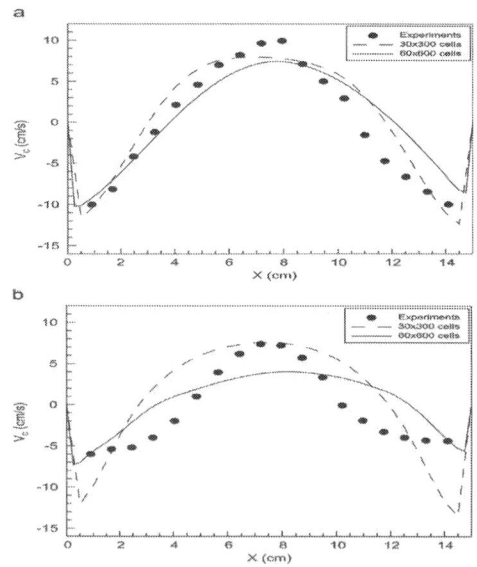

Figure 10: Time-averaged axial liquid velocity versus horizontal position, comparing CFDLib simulations using the Schiller–Naumann drag coefficient model with the BP and BIT models for different grid resolutions to experimental data by Mudde et al. (1997) at heights of (a) H=0.30m and (b) H=0.75m.

Predictions of the gas holdup profile using CFDLib with the Schiller–Naumann drag coefficient model and the BP and BIT models are shown in Fig. 11. Unfortunately, Mudde et al. (1997) did not report experimental gas holdup data and thus, comparisons cannot be made. Although the gas holdup plots in the magnified scale of Fig. 11 (e.g., relative to Fig. 8) show two maxima, the variation in gas holdup is on the order of 1%. Hence $_d$ can be approximated as uniform, indicating that the flow field is extremely homogeneous at this low superficial gas velocity. It should be noted that the homogeneous behavior is only produced when both the BP and BIT models are used. The uniform gas holdup profile at low superficial gas velocity is consistent with observations in the literature (Joshi, 2001).

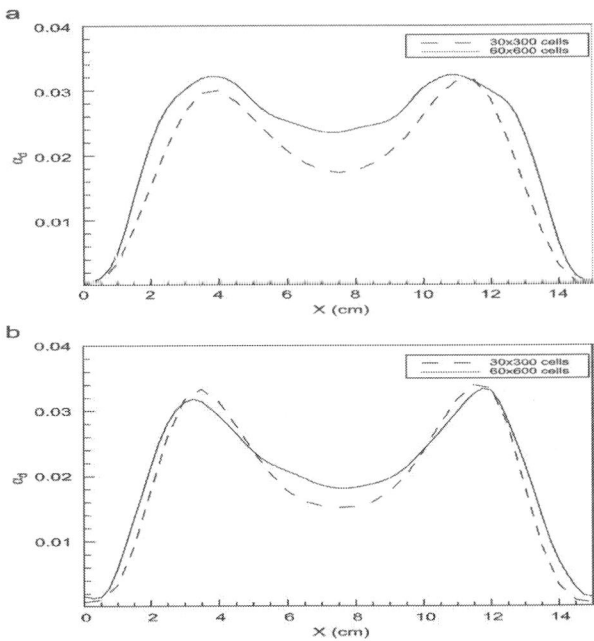

Figure 11: Time-averaged gas holdup versus horizontal position, comparing CFDLib simulations using the Schiller–Naumann drag coefficient model with the BP and BIT models for different grid resolutions at heights of (a) H=0.30m and (b) H=0.75m. Note that the scale is 10 times smaller than previous plots of gas holdup.

Drag Coefficient Model Study

The White drag coefficient model is implemented in CFDLib to study the effect of the drag model on the numerical predictions. Different grid resolutions with and without the BP and BIT models are simulated. Table 2 depicts that for the coarse and medium grid resolutions, where the cell size is larger than the effective bubble diameter (0.5 cm), the simulations converge using the White drag coefficient model with and without the BP model. However, neither simulation with nor without the BP model is able to predict a converged solution for 60×600 cells. A stable time-dependent solution is possible using CFDLib with the White drag coefficient model and both the BP and BIT models. Fig. 12 compares the experimental axial liquid velocity data of Mudde et al. (1997) to the CFDLib simulations using the Schiller–Naumann and White drag coefficient models with 60×600 grid cells and the BP and BIT models. The White model matches the experimental data better than the Schiller–Naumann model at both axial heights. The better prediction by the White model can be attributed to the drag coefficient being a function of bubble Reynolds number instead of a constant value when the bubble Reynolds number exceeds 1000; the latter is the case when the Schiller–Naumann model is used (Eq. (16)). Time-averaged gas holdup distributions are more homogeneous across the bubble column, especially towards the centerline and closer to the walls, when the White model is used as shown in Fig. 13. The larger homogeneous gas holdup region predicted by the White model can be associated with the higher axial liquid velocity predictions (Fig. 12) as compared to the Schiller–Naumann drag coefficient model. Thus, the White drag coefficient model is better suited for low gas velocity flows (which are typically homogeneous flows) where the bubble Reynolds number may moderately exceed a value of 1000.

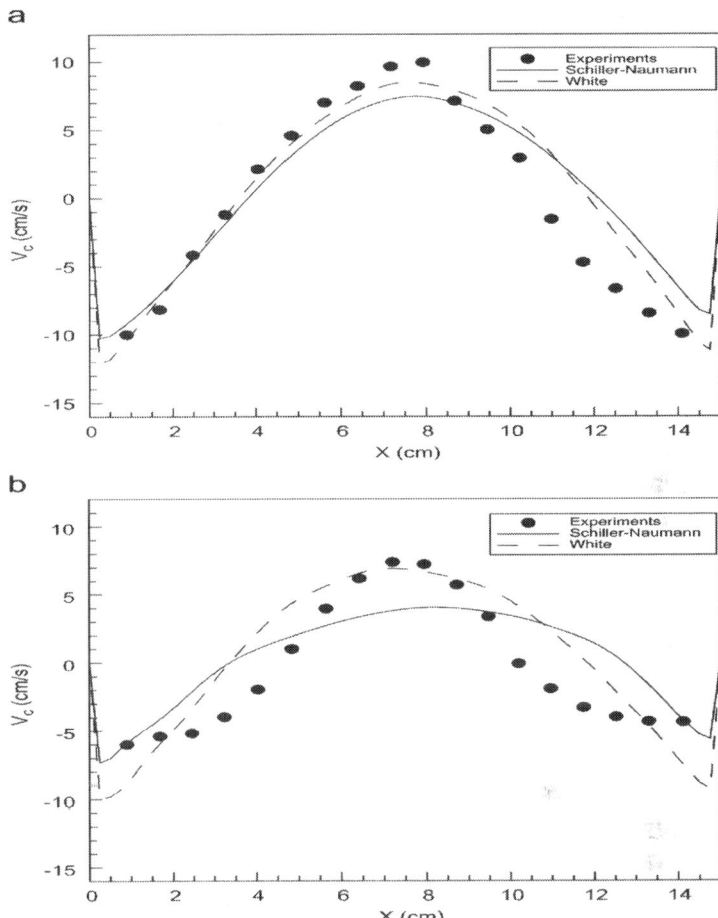

Figure 12: Time-averaged axial liquid velocity versus horizontal position, comparing CFDLib simulations using the Schiller–Naumann and White drag coefficient models for 60×600 grid cells with the BP and BIT models to experimental data by Mudde et al. (1997) at heights of (a) H=0.30m and (b) H=0.75m.

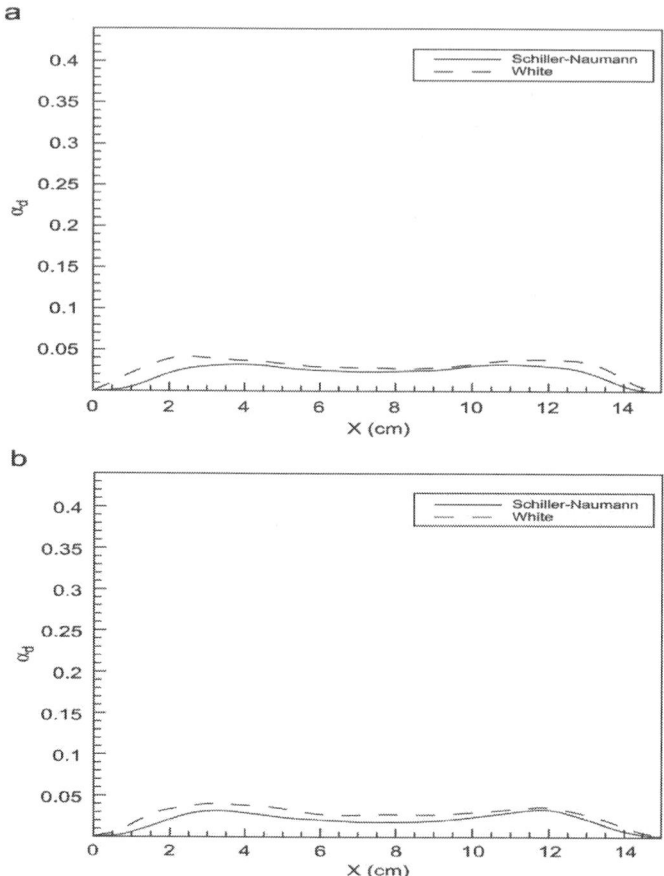

Figure 13: Time-averaged gas holdup versus horizontal position, comparing CFDLib simulations using the Schiller–Naumann and White drag coefficient models for 60×600 grid cells with the BP and BIT models to experimental data by Mudde et al. (1997) at heights of (a) H=0.30m and (b) H=0.75m.

NUMERICAL STABILITY

The work presented herein demonstrates convergence issues with increasing grid resolution when the computational cells are smaller than the effective bubble diameter, irrespective of the superficial gas velocity. It has been shown in the literature (Ransom and Hicks,

1984) that the two-phase flow equations with a single pressure model are known to be ill-posed because they do not result in a system of hyperbolic equations but rather have complex characteristics values. Thus, the ill-posed nature of the single-pressure model is an unbounded instability that is unphysical when using only a hydrostatic assumption. Ransom and Hicks (1984) derive well-posed two-phase flow equations using a two-pressure model that form a hyperbolic system of equations, which are also stable in the sense of von Neumann (that is, without viscosity). Kashiwa and Rauenzahn (1994) reported that ill-posed initial-value problems for the two-phase flow equations with a single-pressure model can be resolved by the use of a properly hyperbolic operator as a means of removing destabilizing truncation errors due to forward-time differencing of the state variables in order to obtain a well-behaved numerical problem. Stewart, 1979 and Stewart, 1981 stated that the ill-posed two-fluid equations can be solved numerically without non-convergence and give well-behaved results if an interfacial drag is present and mesh spacing is not unreasonably fine, which further substantiates the findings in the present work. However, the present work also demonstrates that the implementation of the BP and BIT models into the two-phase flow equations for the low superficial gas velocity case provides a stable time-dependent solution at finer mesh spacing relative to the effective bubble diameter. A linear stability analysis of these equations has been presented by Monahan and Fox (2007) to demonstrate the effects of adding contributions to the pressure term that include gas holdup, slip velocity, and effective viscosity (refer to Section 2.5). Therefore, a more stable numerical solution can be achieved for finer grid resolution, up to a limit, where the cell size is not less than one-half of the effective bubble diameter.

CONCLUSIONS

The gas–liquid flow dynamics in a bubble column were simulated using CFDLib in a two- dimensional Cartesian coordinate system with different grid resolutions to test the effect of using a bubble pressure model and two different drag coefficient models. Simulations at two axial heights above the inlet were compared to the experimental measurements of Rampure et al. (2003) and Mudde et al. (1997) for high and low superficial gas velocities, respectively. From previous FLUENT simulations (Law et al., 2006), finer grid resolutions were not possible if the cell size was smaller than the bubble diameter for the high superficial gas velocity case. On the other hand, the CFDLib simulations produced grid-independent solutions at finer grid resolutions but still failed at a grid resolution of 80×800 cells. Overall, the CFDLib solutions were more physical compared with the FLUENT simulations. The White drag coefficient model predicted homogeneous gas holdup across the bubble column without using the BP model. The Schiller–Naumann model predicted off-center regions of high gas holdup at lower axial liquid heights. The Schiller–Naumann and White models compared qualitatively well with experiments at lower and higher axial liquid heights, respectively. For the low superficial gas velocity case, the higher grid resolution simulation converged when both the BP and BIT models were used in CFDLib. The BP and BIT models provide numerical stability to the simulation, so that cell sizes smaller than the effective bubble diameter can be used. The gas holdup profile was shown to be uniform for the low superficial gas velocity, as expected. The White drag coefficient model predicted the time-averaged axial liquid velocity qualitatively better than the Schiller–Naumann model. In summary, the CFDLib simulations provided physical and grid-independent predictions with the experiments as the grid resolution became finer. It was shown that the BP model did not affect the flow field predictions for the high superficial gas velocity case. Furthermore, the drag coefficient models appear to be sensitive to column location, where the White model works better at low axial heights and the Schiller–Naumann model works better at high axial heights. Finally, for the low superficial gas velocity case, the BP model was needed to model the pressure field and the BIT model added numerical stability to the system. The drag model had minimal effect on the flow predictions for a homogeneous flow field.

ACKNOWLEDGMENTS

This work is supported by the Cooperative State Research, Education, and Extension Service, U.S. Department of Agriculture (USDA), under Agreement No. 2004-34188-15067. Any opinions, findings, conclusions, or recommendations expressed herein are those of the authors and do not necessarily reflect the views of the USDA. Hardware and technical support from the High Performance Computing Center at Iowa State University is also acknowledged. The authors would also like to thank Dr. Bryan (Bucky) A. Kashiwa at Los Alamos National Laboratory for his support and assistance with CFDLib.

REFERENCES

1.   Biesheuvel, A., Gorissen, W.C.M., 1990. Void fraction disturbances in a uniform bubbly fluid. International Journal of Multiphase Flow 16, 211–231.

2.   Delnoij, E., Lammers, F.A., Kuipers, J.A.M., van Swaaij, W.P.M., 1997a. Dynamic simulation of dispersed gas–liquid two-phase flow using a discrete bubble model.

3.   Chemical Engineering Science 52 (9), 1429–1458.

4.   Delnoij, E., Kuipers, J.A.M., van Swaaij, W.P.M., 1997b. Dynamic simulation of gas–liquid two-phase flow: effect of column aspect ratio on the flow structure.

5.   Chemical Engineering Science 52 (21/22), 3759–3772.

6.   Delnoij, E., Kuipers, J.A.M., van Swaaij, W.P.M., 1997c. Computational fluid dynamics applied to gas–liquid contactors. Chemical Engineering Science 52 (21/22), 3623–3638.

7.   Hirt, C.W., Amsden, A.A., Cook, J.L., 1974. An arbitrary Lagrangian–Eulerian computing method for all flow speeds. Journal of Computational Physics 14, 227–253.

8.   Joshi, J.B., 2001. Computational flow modeling and design of bubble column reactors. Chemical Engineering Science 56 (21–22), 5893–5933.

9.   Joshi, J.B., Vitankar, V.S., Kulkarni, A.A., Dhotre, M.T., Ekambara, K., 2002. Coherent flow structures in bubble column reactors. Chemical Engineering Science 57 (16), 3157–3183.

10. Kashiwa, B.A., Rauenzahn, R.M., 1994. A multimaterial formalism. LA-UR-94-771.

11. Kashiwa, B.A., VanderHeyden, W.B., 2000. Toward a general theory for multiphase turbulence. LA-13773-MS Report.

12. Lance, M., Bataille, J., 1991. Turbulence in the liquid phase of a uniform bubbly air–water flow. Journal of Fluid Mechanics 222, 95–118.

13. Launder, B.E., Spalding, D.B., 1974. The numerical computation of turbulent flows. Computer Methods in Applied Mechanical Engineering 3, 269–289.

14. Law, D., Battaglia, F., Heindel, T.J., 2006. Numerical simulations of gas–liquid flow dynamics in bubble columns. In: Proceedings of the ASME Fluids Engineering Division, IMECE2006-13544, Chicago, IL.

15. Lin, T.-J., Reese, J., Hong, T., Fan, L.-S., 1996. Quantitative analysis and computation of two-dimensional bubble columns. A.I.Ch.E. Journal 42, 301–318.

16. Mizukami, M., Parthasarathy, R.N., Faeth, G.M., 1992. Particle-generated turbulence in homogeneous dilute dispersed flows. International Journal of Multiphase Flow 18, 397–412.

17. Monahan, S.M., Fox, R.O., 2007. Linear stability analysis of a two-fluid model for air–water bubble columns. Chemical Engineering Science 62, 3159–3177.

18. Monahan, S.M., Vitankar, V.S., Fox, R.O., 2005. CFD predictions for flow-regime transitions in bubble columns. A.I.Ch.E. Journal 51, 1897–1923.

19. Mudde, R.F., Lee, D.J., Reese, J., Fan, L.-S., 1997. Role of coherent structures on Reynolds stresses in a 2-D bubble column. A.I.Ch.E. Journal 43, 913–926.

20. Pan, Y., Dudukovic, M.P., 2000. Numerical investigation of gas-driven flow in 2-D bubble columns. A.I.Ch.E. Journal 46, 434–449.

21. Parthasarathy, R.N., Faeth, G.M., 1990a. Turbulence modulation in homogeneous dilute particle-lade flows. Journal of Fluid Mechanics 220, 485–514.

22. Parthasarathy, R.N., Faeth, G.M., 1990b. Turbulent dispersion of particles in selfgenerated homogenous turbulence. Journal of Fluid Mechanics 220, 515–537.

23. Rafique, M., Dudukovic, M.P., 2006. Influence of different closures on the hydrodynamics of bubble column flows. Chemical Engineering Communications 193, 1–23.

24. Rampure, R.M., Buwa, V.V., Ranade, V.V., 2003. Modeling of gas–liquid/ gas–liquid–solid flows in bubble columns: experiments and CFD simulations. The Canadian Journal of Chemical Engineering 81, 692–706.

25. Ransom, V.I I., Hicks, D.L., 1984. Hyperbolic two-pressure models for two-phase flow. Journal of Computational Physics 53, 124–151.

26. Sanyal, J., Vasquez, S., Roy, S., Dudukovic, M.P., 1999. Numerical simulation of gas–liquid dynamics in cylindrical bubble column reactors. Chemical Engineering Science 54, 5071–5083.

27. Sato, Y., Sadatomi, M., Sekoguchi, K., 1981. Momentum and heat transfer in twophase bubble flow: I. International Journal of Multiphase Flow 7, 167–177.

28. Schiller, L., Naumann, A., 1933. Uber die grundlegenden berechnungen bei der schwerkraftaufbereitung. Zeitung des vereins deutscher ingenieure, 77–318.

29. Sokolichin, A., Eigenberger, G., 1994. Gas–liquid flow in bubble columns and loop reactors: I. Detailed modeling and

numerical simulation. Chemical Engineering Science 49, 5735–5746.

30. Spelt, P.D.M., Sangani, A., 1998. Properties and averaged equations for flows of bubbly liquids. Applied Science Reserve 58, 337–386.

31. Stewart, H.B., 1979. Stability of two-phase flow calculation using two-fluid models. Journal of Computational Physics 33, 259–770.

32. Stewart, H.B., 1981. Fractional step methods for thermohydraulic calculation. Journal of Computational Physics 40, 77–90.

33. White, F.M., 1974. Viscous Fluid Flow. McGraw-Hill, New York.

Chapter 2

Fracturing Fluids

Carl Montgomery[1]

[1]NSI Technologies, and Tulsa, Oklahoma, USA

ABSTRACT

When fracturing, viscosity play a major role in providing sufficient fracture width to insure proppant entrance into the fracture, carrying the proppant from the wellbore to the fracture tip, generating a desired net pressure to control height growth and providing fluid loss control. The fluid used to generate the desired viscosity must be safe to handle, environmentally friendly, non-damaging to the fracture conductivity and to the reservoir permeability, easy to mix, inexpensive and able to control fluid loss. This is a very demanding list of requirements that has been recognized since the beginning of Hydraulic fracturing. This paper describes the history of fracturing fluids, the types of fracturing fluids used, the engineering requirement of a good fracturing fluid, how viscosity is measured and what the limitations of the engineering design parameters are.

INTRODUCTION

The selection of a proper fracturing fluid is all about choices. It begins with choosing the pad volume where one must consider what and how much pad is required to create the desired fracture geometry. This is followed by choosing how much viscosity the fluid needs to have to:

- Provide sufficient fracture width to insure proppant entrance into the fracture.

- Provide a desired net pressure to either treat some desired height growth or prevent breaking out into some undesirable zone for example water.

- Provide carrying capability to transport proppant from the wellbore to the fracture tip.

- Control fluid loss. In cases where a gel filter cake cannot form the fracturing fluid viscosity (i.e. C_l) may be the main mechanism for fluid loss control.

This choice system continues when it comes to selecting the appropriate fluid system for a propped or acid frac treatment. The considerations include:

- Safe – The fluid should expose the on-site personnel to a minimal danger.

- Environmentally Friendly – The composition of the fluid should be as "green" as possible.

- Breaker – The fluid must "break" to a low viscosity so that it can flow back and allow clean-up of the fracture.

- Cost Effective – The fluid must be economical and not drive the treatment cost to an unacceptable level.
- Compatibility – The fluid must not interact and caused damage with the formation mineralogy and/or formation fluids.
- Clean-up – The fluid should not damage the fracture conductive of the fracture or, to prevent water blocks, change the relative permeability of the formation. This becomes very important in low pressure wells or wells that produce very dry gas.
- Easy to Mix – The fluid system must be easy to mix even under very adverse conditions.
- Fluid Loss – The fluid need to help control fluid loss. An ideal fluid should have fluid loss flexibility.

In summary an ideal fracturing fluid would be one that would have an easily measured controllable viscosity, controllable fluid loss characteristics, would not damage the fracture or interact with the formation fluid, would be completely harmless and inert and cost less the $4.00 US/ gallon. Unfortunately this is currently not possible so compromises have to be made. Typically cost is the driving force and chooses are made which can be disastrous to the PI of the well.

Of these factors the fluid viscosity is the major fluid related parameter for fracture design. However, how much viscosity needed is often overrestimated. Excessive viscosity increases costs, raises treating pressure which may cause undesired height growth, and can reduce fracture conductivity since many of the chemicals used to increase viscosity leave residue which damages the proppant permeability.

The need for a precise value of viscosity is also over engineered. This can be seen from the basic equations where treating pressure, and thus fracture width, is proportional to viscosity raised to the ¼ power (for a Newtonian fluid).

$$P_{net} \propto \frac{E'^{3/4}}{H}[\mu Q L]^{1/4} + P_{Tip}$$

Thus a 100% error in viscosity results in an error of about 19% in calculating fracture width. This error would, of course, lead to an error in the fluid volume requirements for a particular job. However, further assuming that 1/2 of the fracturing fluid leaks off to the formation reduces the 19% error in width to only a 9.5% error in fluid volume requirements. While such an error is not desirable it does illustrate that precise viscosity data is not a requirement for treatment design which is fortunate since the measurement of the viscosity of fracturing fluids is such a difficult task. This complexity combined with multiple methods for testing and reporting viscosity data makes the selection of precise values virtually impossible.

There are several types of fracturing fluids and a wide and confusing range of fluid additives. The types of fluids include:

- Water based fluids
- Oil based fluids
- Energized fluids
- Multi-phase emulsions
- Acid Fluids

The additives include:

- Gelling agents
- Crosslinkers
- Breakers
- Fluid loss additives
- Bactericides
- Surfactants and Non-emulsifing agents
- Clay control Additives.

HISTORY

The fracturing fluids that were used in the first experimental treatments were composed of gasoline gelled with Palm Oil and crosslinked with Naphthenic Acid. This technology was developed during the Second World War and is commonly referred to as

Nalpalm. Because of the hazards associated with this fluid and its relatively high cost work was done to develop safer fluids where the base fluid was water. The vast majority of fracturing fluids used today use water as the base fluid. Generally, the components that make up crosslinked fracturing fluids include a polymer, buffer, gel stabilizer or breaker and a crosslinker. Each of these components is critical to the development of the desired fracturing fluid properties. The role of polymers in fracturing fluids is to provide fracture width, to suspend proppants, to help provide fracture width, to help control fluid loss to the formation, and to reduce friction pressure in the tubular goods. Guar gum and cellulosic derivatives are the most common types of polymers used in fracturing fluids. The first patent (US Patent 3058909) on guar crosslinked by borate was issued to Loyd Kern with Sinclair (later ARCO) on October 16, 1962. Metal-based crosslinking agents developed by DuPont for plastic explosive applications were found to be useful for manufacturing fracturing fluids for high temperature applications[2]. Cellulosic derivatives are residue-free and thus help minimize fracturing fluid damage to the formation and are widely used in Frac and Pack applications. The cellulosic derivatives are difficult to disperse because of their rapid rate of hydration. Guar gum and its derivatives are easily dispersed but produce some residue when broken. Strong oxidizing agents such as Sodium or Ammonium persulfate are added to the fracturing fluids to break the polymer as it reaches temperature. The first patent (US Patent 3163219) on borate gel breakers was issued to Tom Perkins, also with Sinclair, on December 29, 1964.

Buffers are used in conjunction with polymers so that the optimal pH for polymer hydration can be attained. When the optimal pH is reached, the maximal viscosity yield from the polymer is obtained. The most common example of fracturing fluid buffers is a weak-acid/weak-base blend, whose ratios can be adjusted so that the desired ph is reached. Some of these buffers dissolve slowly allowing the crosslinking reaction to be delayed.

Gel stabilizers are added to polymer solutions to inhibit chemical degradation. Examples of gel stabilizers used in fracturing fluids include methanol, TriEthanol Amine (TEA) and various inorganic sulfur compounds. Other stabilizers are useful in inhibiting the chemical degradation process, but many interfere with the mechanism of crosslinking. The TEA and sulfur containing stabilizers possess an advantage over methanol, which is flammable, toxic, expensive and can cause poisoning of reactor tower catalists.

There has been a huge volume of work done on fracturing fluids and their components. If a search is done on One Petro (http:// www.onepetro.org) using "Fracturing Fluids" as the search item over 15,000 hits will result. Just using one of the main gelling agents used to manufacture water based fracturing fluid "Guar" results in over 400 hits. There are several good references [3,4,5,6] that discuss the current state of the art for fracturing fluids if the reader is interested in a more in depth study of fracturing fluids.

Another issue that has recently come to the forefront of fracturing fluids is their threat to the environment through the contamination of the groundwater. George King put it very elegantly in his JPT article[7] where he says "The use of horizontal wells and hydraulic fracturing is so effective that it has been called "disruptive". That is, it threatens the profitability and continued development of other energy sources, such as wind and solar, because it is much less expensive and far more reliable." The internal Apache article[8] that George wrote has 204 references on the subject. Table 1,2,3 provides a summary of all the various chemicals used to make Hydraulic Fracturing fluids along with a degree of hazard rating from both the US Department of Transportation and the European Union Poison Class rating. There certainly are several of these chemicals that one must take care with when handling at their full concentrations but when used to manufacture fracturing fluids the concentrations are very dilute and pose very low hazards.

Table 1: A summary of the various chemicals used to make Hydraulic Fracturing fluids along with a degree of hazard rating. Modified from" www. http://fracfocus.org/chemical-use/what-chemicals-are-used"

Chemical Name	CAS Number	Chemical Purpose	Product Function	Hazard Rating[1]
Hydrochloric Acid HCl	007647-01-0	Removes acid soluble minerals and weakens the rock to allow lower fracture iniciation pressures.	Acid	4*,8**
Glutaraldehyde $C_5H_{8O}2$	000111-30-8	Eliminates bacteria in the water to prevent frac polymer premature breakdown and well souring	Biocide	3*,6**
Quaternary Ammonium Chloride Compounds	63393-96-4	Clay Control Agents	Biocides and Clay Stabilizers	3**
Tetrakis Hydroxymethyl-Phosphonium Sulfate $C_8H_{24}O_8P_2.S_O4$	055566-30-8	Eliminates bacteria in the water to prevent frac polymer premature breakdown and well souring	Biocide	NR
Ammonium Persulfate $(NH_4)_2S_{2O}8$	007727-54-0	Breaks the polymer that is used to create the fracturing fluid	Breaker	4*,5**
Sodium Chloride NaCl	007647-14-5	Product Stabilizer	Breaker	NR
Magnesium Peroxide Mg_O2	1335-26-8	Delays the breakdown of the fracturing fluid gelling agent	Breaker	5**
Magnesium Oxide MgO	1309-48-4	Delays the cross linking of the fracturing fluid gelling agent	Buffer	4*
Calcium Chloride CaC_l2	10043-52-4	Product Stabilizer and Freeze Protection	Buffer	NR
Ammonium Chloride NH_4Cl	012125-02-9	Clay Stabilizer – Compatible with Mud Acid	Clay Stabilizer	4*,9**
Choline Chloride $[HOCH_2CH_2N^+(CH_3)_3]$ C	67-48-1	Prevents clays from swelling or migrating	Clay Stabilizer	5*
Potassium chloride KCl	007447-40-7	Prevents clays from swelling or migrating	Clay Stabilizer	5*,5**
Tetramethyl ammonium chloride $(CH_3)_4NCl$	000075-57-0	Prevents clays from swelling or migrating	Clay Stabilizer	3*,6**
Sodium Chloride NaCl	007647-14-5			NR

Isopropanol $CH_3CH(OH)C_H3$	000067-63-0	Winterizing agent	Winterizing agent and Surface Tension Reduction	3**
Methanol CH_3OH	000067-56-1	Winterizing agent	Winterizing agent	3*, 3**
Formic Acid HCOOH	000064-18-6	pH adjustment	pH adjustment	4*,8**
Acetaldehyde CH_3CHO	000075-07-0	Prevents the corrosion of the pipe	Corrosion Inhibitor	4*,3**
Hydrotreated Light Petroleum Distillate	064742-47-8	Carrier fluid for gelling agents, friction reducers and crosslinkers	Carrier fluid and fluid loss control	3**
Potassium Metaborate KB_O2	013709-94-9	Crosslinker for borate crosslinked fluids	Crosslinker	3*
Triethanolamine (TEA) $N(CH_2CH_2OH)3$	102-71-6	Maintains fluid viscosity as temperature increases	Fluid Stabilizer	5*,3**
Sodium Tetraborate $Na_2B_{4O}7$	001330-43-4	Crosslinker for borate crosslinked fluids	Crosslinker	4*
Boric Acid H_3B_O3	13343-35-3	Crosslinker for borate crosslinked fluids	Crosslinker	4*
Chelated Zirconium		Crosslinker for High Temperature or low pH Fluids	Crosslinker	
Zirconium oxychloride $ZrCl_2O$	7699-43-6	Inorganic Clay Stabilizer	Clay Stabilizer	4*
Ethylene Glycol OCH_2CH_2OH	000107-21-1	Product stabilizer and / or winterizing agent.	Winterizing Agent	4*
Methanol CH_3OH	000067-56-1	Surface Tension Reduction and / or winterizing agent.	Fluid Recovery and Winterizing Agent	3*,3**
Ethanol C_2H_5OH	000064-17-5	Product stabilizer and / or winterizing agent.	Fluid Recovery and Winterizing Agent	3**
Polyacrylamide $(C_3H_5NO)n$	009003-05-8	"Slicks" the water to minimize friction	Friction Reducer	5*
Guar Gum and its derivatives HPG, CMHPG	009000-30-0	Thickens the water in order to suspend the proppant and reduce friction	Gelling Agents	NR
Derivatives of cellulose - HEC, CMHEC $R(n)OCH_2COONa$	9004-34-6 9004-32-4	Thickens the water in order to suspend the proppant and reduce friction	Gelling Agents	NR
Xanthan gum	11138-66-2	Thickens Acid in order to control fluid loss	Gelling Agent	NR

Citric Acid $(HOOCCH_2)_2C(OH)COOH$	000077-92-9	Prevents precipitation of metal oxides	Iron Control	5*,8**
Acetic Acid CH_3COOH	000064-19-7	Prevents precipitation of metal oxides and pH control	Iron Control and pH Adjustment	4*,8**
Thioglycolic Acid $HSCH_2COOH$	000068-11-1	Prevents precipitation of metal oxides	Iron Control	3*,8**
Sodium Erythorbate $C_6H_7O_6$. Na	006381-77-7	Prevents precipitation of metal oxides	Iron Control	NR
Lauryl Sulfate and its Derivatives $C_{12}H_{25}OSO_2ONa$	000151-21-3	Used to prevent the formation of emulsions in the reservoir and to improve fluid recovery	Non-Emulsifier and Surfactants	4*
Sodium Hydroxide NaOH	001310-73-2	Adjusts the pH of fluid to initiate the effectiveness of other components, such as crosslinkers	pH Adjusting Agent	4*,8**
Potassium Hydroxide KOH	001310-58-3	Adjusts the pH of fluid to initiate the effectiveness of other components, such as crosslinkers	pH Adjusting Agent	2*,8**
Sodium Carbonate Na_2C_O3	000497-19-8	Adjusts the pH of fluid to maintains the effectiveness of other components, such as crosslinkers	pH Adjusting Agent	5*,5**
Potassium Carbonate K_2C_O3	000584-08-7	Adjusts the pH of fluid to maintains the effectiveness of other components, such as crosslinkers	pH Adjusting Agent	4*
Sodium Acrylate and Copolymers of Acrylamide $C_3H_3O_2$. Na	007446-81-3	Prevents scale deposits in the pipe or in the fracture	Scale Inhibitor	NR
Sodium Polycarboxylate	N/A	Prevents scale deposits in the pipe	Scale Inhibitor	
Phosphonic Acid Salt	N/A	Prevents scale deposits in the pipe	Scale Inhibitor	
Naphthalene $C_{10H}8$	000091-20-3	Carrier fluid for the active surfactant ingredients	Surfactant	3*,4**
Ethylene glycol mono-butyl ether – EGMBE $C_4H_9OCH_2CH_2OH$	000111-76-2	Surface Tension Reduction for Fluid Recovery	Surfactant	4*, 6**

[i] - 1 – Hazard Rating – An attempt was made to rate the hazard associated with each of the chemicals listed. The first number with the single * is the Poison Hazard as defined by the EU/Swiss Poison Class while the second number with the double ** is the transportation Hazard as defined by the US Department of Transportation (DOT). If a NR is present in the box no rating was found and the substance was normally non-hazardous.

[ii] - * EU/Swiss Poison Class

Table 2: A summary of the various chemicals used to make Hydraulic Fracturing fluids along with a degree of hazard rating. Modified from" www. http://fracfocus.org/chemical-use/what-chemicals-are-used"

Class	Lethal Dose (mg/kg)
1	0 to 5
1S	0 to 5, also teratogenic or carcinogenis
2	5 to 50
3	50 to 500
4	500 to 2000
5	2000 to 5000
5S	2000 to 5000, an unrestricted self-service product

[i] - ** DOT Transportation Hazard Classes

Table 3: A summary of the various chemicals used to make Hydraulic Fracturing fluids along with a degree of hazard rating. Modified from" www. http://fracfocus.org/chemical-use/what-chemicals-are-used"

Class	
1	Explosives
2	Compressed Gases
3	Flammable and Combustible Liquids
4	Flammable Solids
5	Oxidizers and Organic Peroxides

6	Poisonous/Toxic Materials
7	Radioactive Materials
8	Corrosive Materials
9	Miscellaneous Hazardous Materials

Additional Hazard Identification Resources

http://fracfocus.org/welcome - The Ground Water Protection Council and the Interstate Oil and Gas Compact Commission developed this web site to provide public access to chemicals used in the hydraulic fracturing process and provides a record of the chemicals used in wells in a number of different stated in the United States. At the time of this writing the site had records on over 34,000 wells.

http://www.osha.gov/chemicaldata/ - This United States Department of Labor website proves a OSHA (Occupational Safety and Health Administration) Occupational Chemical Database for most of the chemicals used by industry. The database can be searched by either Chemical Name or CAS Number.

http://ull.chemistry.uakron.edu/erd/ - The Department of Chemistry at the University of Akron developed this website to provide a database composed of over 30,000 hazardous chemicals made up of information provided by a number of different published references.

http://www.epa.gov/chemfact/ - This United States Environmental Protection Agency website provides OPPT Chemical Fact Sheets on selected chemicals that may be present in the environment in an ASCII text or Adobe PDF format along with access to other EPA databases.

TYPES OF FRACTURING FLUIDS

Table 4 provides a qualitative listing of the desirable and undesirable aspects of most fluid systems available today. As one studies the table it is interesting to note that there is "no magic bullet". The qualitative score is close to the same for each fluid and each fluid has its advantages and disadvantages. This means that the final decision is up to the design engineer as to what is best for his reservoir. The different types of fluid systems are outlined below. A description of all the different components used to manufacture the fluids is provided in Side Bar 1.

Water Frac is composed of water, a clay control agent and a friction reducer. Sometimes a water recovery agent (WRA) is added to try and reduce any relative permeability or water block effects. The main advantage of using a "Water Frac" is the low cost, ease of mixing and ability to recover and reuse the water. The main disadvantage is the low viscosity which results in a narrow fracture width. Because the viscosity is low the main proppant transport mechanism is velocity so water fracs are typically pumped at very high rates (60 to 120 bpm). Fluid loss is controlled by the viscosity of the filtrate which is close to that of water i.e. 1.

Linear Gel is composed of water, a clay control agent and a gelling agent such as Guar, HPG or HEC. Because these gelling agents are susceptible to bacteria growth a bactericide or biostat is also added. Chemical breakers are also added to reduce damage to the proppant pack. WRA's are also sometimes used. The main advantage of a liner gel is its low cost and improved viscosity characteristics. Fluid loss is controlled by a filter cake which builds on the fracture face as the fluid loses fluid to the formation. The main disadvantage is, as with waterfracs, the low viscosity which results in a narrow fracture width. The main disadvantage when compared to a waterfrac is that because the returned water has residual breaker the water is not reusable.

Table 4: Qualitative Fluid Selection Chart

Fluid System	Prop Pack KtW	Low Pump Pressure	VISCOSITY			Breaking	Compat bility		Fluid Loss	Ease of Mixing	Cost	Safety and Environmentally Friendly	Total
			Prop-Transport	Stable	Life		Formation Fluid	Fluid Recovery					
Water Frac[1]	5	5	1	3	3	5	3	4	1	5	5	4	44
Linear Gel[2]	3	5	3	3	3	4	3	4	2	5	4	5	44
Linear Gell	5	5	3	3	3	4	3	4	2	5	4	5	46
Borate X-Link[2]	3	3	5	5	5	3	4	3	5	4	3	5	48
Delayed Borate X-Link[2]	3	3	5	5	5	3	4	3	5	3	3	5	47
Delayed Metallic X-Link[4]	3	3	5	2	2	3	4	3	5	3	3	4	40
Delayed Metallic X-Links	3	3	5	2	2	3	4	3	5	3	3	4	40
VES[6]	5	3	5	4	4	2	1	3	2	2	1	5	37
Nitrogen Foam	5	2	5	3	3	5	4	4	5	2	1	3	42
CO_2 Foams	5	2	5	3	3	5	4	5	5	2	1	2	42
Gelled Propane	5	3	4	4	3	4	5	4	4	2	1	1	40
Poly Emulsions (K1)	4	1	5	5	5	4	4	3	5	2	3	2	43
Lease Crude	2	3	2	5	5	5	5	3	2	5	5	1	43
Gelled 011[7]	2	3	4	4	4	4	3	3	3	4	3	1	38

Qualitative Rate 1 to 5 where 1 is poor, 3 is moderate and 5 is excellent

[1]Uses Polyacrylamide (PAA) as a Friction Reducer

[2]Uses Guar, HydroxyPropyl Guar (HPG) or CarboxyMethylHydroxyPropyl Guar (CMHPG) as gelling agent

[3]Uses HydroxyEthyl Cellulose (HEC) or CarboxyMethyHydroxyEthyl Cellulose (CMHEC) as gelling agent

[4]Uses Titinium or Zirconium Crosslinkers for Guar, HPG, and CMHPG gelling agents

[5]Uses Titinium or Zirconium Crosslinkers for CMHEC gelling agents

[6]Uses a ViscoElastic Surfactant system as the gelling agent

[7]Uses a Phosphate Ester crosslinked with an Aluminum Salt and acitivated with a Base

Crosslinked Gels are composed of the same materials as a linear gel with the addition of a crosslinker which increases the viscosity of the linear gel from less than 50 cps into the 100's or 1000's of cps range. The higher viscosity increases the fracture width so it can accept higher concentrations of proppant, reduces the fluid loss to improve fluid efficiency, improves proppant transport and reduces the friction pressure. This crosslinking also increases the elasticity and proppant transport capability of the fluid. Fluid loss is controlled by a filter cake which builds on the fracture face as the fluid loses fluid to the formation. A full description of the types of crosslinkers used, the chemistry and the mechanism of crosslinking is provided in the companion paper on fracturing fluid components.

Oil Based Fluids are used on water-sensitive formations that may experience significant damage from contact with water based fluids. The first frac fluid used to fracture a well used gasoline at the base fluid, Palm Oil as the gelling agent and Naphthenic Acid as the crosslinker i.e. Napalm. Although some crude oils have particulate which could build a filter cake, fluid loss is generally considered to be "Viscosity- Controlled – i.e. C-II". There are some disadvantages in using gelled oils. Gelling problems can occur when using high viscosity crude oils or crude oils which contain a lot of naturally occurring surfactants. When using refined oils such as diesel the cost is very high and the oil must be collected at the refinery before any additives such as pour point depressants, engine cleaning surfactants etc. are added. Also there are greater concerns regarding personnel safety and environmental impact, as compared to most water-fluids.

Foam/PolyEmulsions are fluids that are composed of a material that is not miscible with water. This could be Nitrogen, Carbon dioxide or a hydrocarbon such as Propane, diesel or condensate. These fluids are very clean, have very good fluid loss control, provide excellent proppant transport and break easily simply via gravity separation. PolyEmulsions are formed by emulsifying a hydrocarbon such as Condensate or Diesel with water such that the hydrocarbon is the external phase. The viscosity is controlled by varying the hydrocarbon/water ratio. Foams made with Nitrogen or Carbon dioxide is generally 65 to 80% (termed 65 to 80 quality) gas in a water carrying media which contains a surfactant based foaming agent. Sometimes N_2 or CO_2 are added at a lower concentration (20 to 30 quality) to form "Energized Fluids". This is done to reduce the amount of water placed on the formation and to provide additional energy to aid in load recover during the post-frac flow back period. Nitrogen can dissipate into the reservoir quite quickly so fluids energized with N_2 should be flowed back as soon as the fracture is closed. CO_2, under most conditions, is in a dense phase at static down hole conditions (prior to the well being placed on production), so is less susceptible to dissipation. CO_2 does dissolve in crude oil so will act to reduce the crude viscosity which, again, improves cleanup and rapid recovery. When N_2/CO_2 are added is qualities greater than 80 the resulting mixture is termed a mist with a "0" viscosity. This quality is normally not used in fracturing. The main disadvantage of these fluids is safety i.e. pumping a gas at high pressure or in the case of polyemulsions and gelled Propane, pumping a flammable fluid. CO_2 has an additional hazard in that it can cause dry ice plugs as pressure is reduced. These fluids are generally also more expensive and the gases may not be available in remote areas.

CHACTERIZATION OF FRACTURING FLUIDS

Fluid viscosity for treatment design is determined from laboratory tests and is reported in service company literature. The ideal experiment for describing fluid flow in a fracture would be to shear a fluid between two plates which are moving parallel and relative to one another. The shear stress on the fluid equals the drag force on the plates divided by the area of the plates, and has units of stress or pressure (e.g., psi). The shear rate (or velocity gradient) is the relative velocity of the two plates divided by the separation distance between the plates. Shear rate has the units of 1/time (e.g., sec^{-1}). A vertical 7 ft high by 10 1/3 ft long high pressure parallel-plate flow cell, shown in Figure 1, capable of operating to temperatures of 250°F and pressures of 1200 psi is available at the University of Oklahoma[11]. Termed the "Fracturing Fluid Characterization Facility (FFCF)" the laboratory simulator is a very sophisticated; one of a kind unit that utilizes 12 servo-controlled 28" by 28" platens that can dynamically adjust the width of the slot from 0 to 1.25 inches.

Figure 1: University of Oklahoma Parallel Plate Fracturing Fluid Characterization Facility (Courtesy of the University of Oklahoma).

Such an ideal test is not feasible for day to day applications so a rotating "cup and bob" viscometer know as a "Couette" viscometer is used. API standard RP39[12] and ISO 13503-1[13] fully describe the current testing procedures used by the industry. The viscometer uses a rotating cup and a stationary bob with a gap between the two that simulates the fracture. As shown in Figure 2 the rotational speed of the cup imparts a shear rate and the bob measures the shear stress or drag force exerted on the walls of the cup and bob. This is sensed by measuring the torque on the bob. The shear rate is the relative velocity between the stationary bob and the rotating cup divided by the separation gap. Figure 3 shows several commercial rheometers and how they are set up in the field. For a Fann 35 (See Figure 3) equipped with a R1 rotor and a B1 bob and the appropriate spring a rotational speed of 100 RPM represents a shear rate of 170 sec^{-1} and a speed of 300 RPM gives a shear rate of 511 sec^{-1}. The Fann 35, which is manufactured by the Fann Instrument Company http://www.fann.com/, the Model 3530, which is manufactured by Chandler Engineering http://www.chandlerengineering.com/ and the Model 800 8 speed viscometer manufactured by OFI Testing Equipment, Inc. http://www.ofite.com/ are atmospheric rheometers which limits their use to the boiling point of water. The Fann 50, Chandler 5550 and OFI 130-77 viscometer's are equipped with a pressurized cup and bob which can be placed into an oil bath for higher temperature measurements. Fluids, including foam, can be dynamically flowed into the cells so that the fluid can be measured under the shear conditions that it would experience in the well. These rheometers are very rugged reliable instruments but suffer from a phenomenon called the Weissenberg effect when trying to measure crosslinked viscoelastic fluids. It occurs when a spinning rod, like the rotor, is placed into a solution of polymer. Instead of being thrown outward the polymer chains entangle on the rod supporting the bob causing the polymer solution to be drawn up the rod. Figure 4 shows what the Weissenberg effect looks like. As temperature increases and the gel thins the issue goes away to a certain extent and modern rheometers try to control the effect. Overall the effect can result in some very misleading data and care must be taken when very odd looking, unusual data is presented.

The testing problem is compounded in that, as illustrated in Figure 5, many fracturing fluids (particularly crosslinked gels) are not truly fluids. Trying to characterize these materials with a "viscosity" can be very difficult. Fortunately, even for these fluids, temperatures above about 120°F make the behavior more predictable.

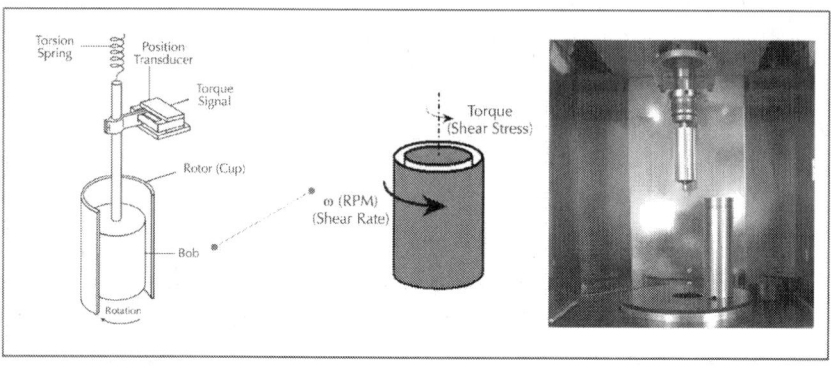

Figure 2: The geometry of a curette "Cup & Bob" Viscometer.

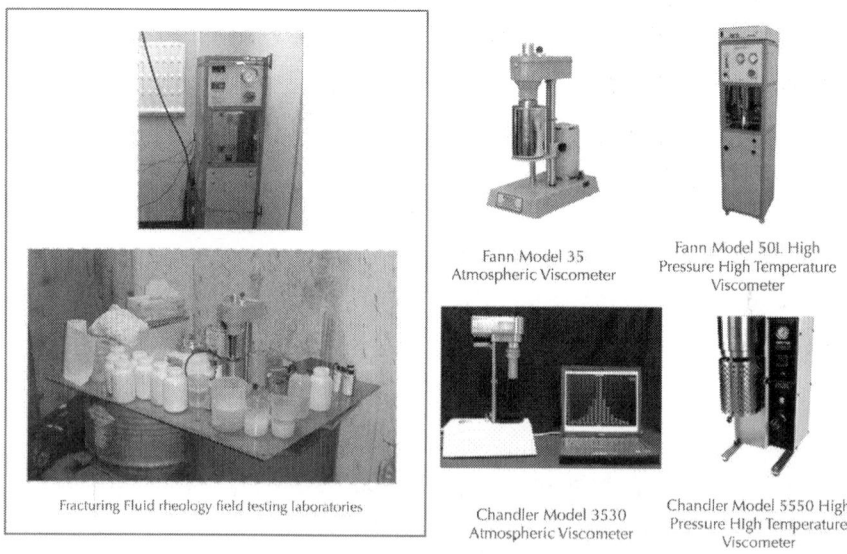

Figure 3: Rheometer's for testing fracturing fluids.

Figure 4: The Weissenberg Effect.

Figure 5: Example of a Complex Dehydrated Cross-linked gel.

RHEOLOGICAL MODELS

The tests described above measure the shear stress generated by specific increasing shear rates (called a ramp), and this data is converted to a "viscosity" value by using a rheological model to describe fluid behavior. Figure 6 shows the three models that are in common use by the oil industry and these are:

- Newtonian Fluid - A Newtonian fluid has a linear relation between shear rate and shear stress and fluid viscosity is the slope of the shear rate versus shear rate data.
- Bingham Plastic - A Bingham Plastic differs from a Newtonian fluid in that a non-zero shear stress called the Plastic Yield Value is required to initiate fluid flow. The slope of the shear rate/shear stress data is labeled Plastic Viscosity and this model is routinely used for cements and many drilling muds.
- Power Law Fluid - This is the most common fluid model used for current fracturing fluids and for this rheological model the shear stress/shear rate data give a linear relation on log-log scales. The slope of this log-log line is denoted by n', and this is labeled the Flow Behavior Index. n'=1 implies a Newtonian fluid; n'>1 is called a shear stiffening fluid; and n'<1 is a shear softening fluid. n' is generally less than 1 for fracturing fluids. The shear stress at a shear rate of "1" is labeled the Consistency Index and is denoted by K'. For real fluids K' and n' change with temperature and time with K' generally decreasing and n' tending toward unity.

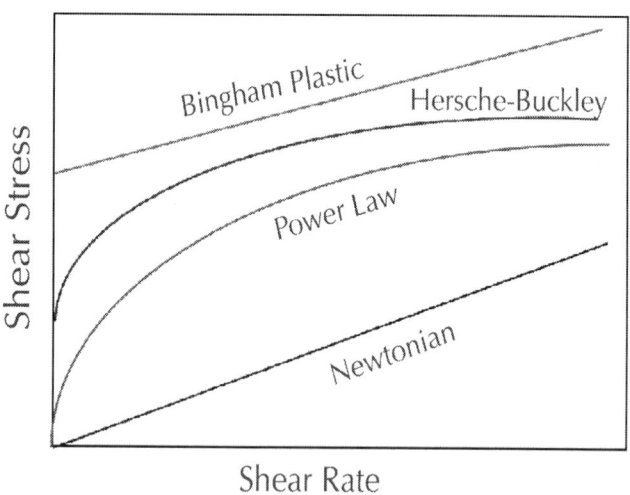

Shear Rate

Figure 6: Rheological Models.

For non-Newtonian fluids (a Power Law fluid being one example) the "apparent viscosity - (u_a) " is used as a shorthand way of characterizing the fluid. Apparent viscosity (u_a) is illustrated in Fig. 7 and is the ratio of shear stress to shear rate - at a particular value of shear rate. Thus a fluids apparent viscosity depends on the shear rate at which the viscosity is measured (or calculated). For a Power Law Fluid with n'<1, the apparent viscosity will decrease with increasing shear rate.

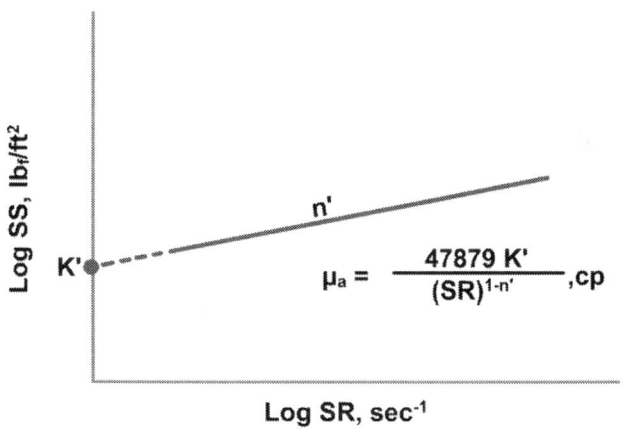

Figure 7: Apparent viscosity using a Power Law Equation.

To determine n' and K' a fluid is placed in a rheometer and sheared at a constant rate while the temperature is brought to equilibrium. Periodically the fluid n' and K' is measured by bringing the shear rate up, holding the rate for a few seconds then increasing the rate again typically over a range of at least 4 shear rates. This is termed a ramp and is typically done every 30 minutes during the fluid test.Figure 8 shows an example of a shear stress vs shear rate set of ramps that was provided by C&A Inc. - http://www.candalab.com/. Note that for each ramp four shear rates where used. The slope of the line is the n' and the intercept at a 0 shear rate is the K'. Using this information an apparent viscosity for any shear rate can be calculated with the following equation.

$\mu a = 448000\ K\mathstrut_{}(SR)1-n'$

Where µa = Apparent viscosity in cps

K' = the Consistency Index in (lbf/ft²/sec)

n' = flow behavior index

SR = Shear Rate in Sec $^{-1}$

Service company literature reports viscosity at different shear rates (usually 170 or 511 sec^{-1}) and the shear rate in a fracture can be as low as 30 to 40 sec^{-1}. The example shows that the identical fluid might be reported by one company to have a viscosity of 300 cp (170 sec^{-1}), by another to have 200 cp (511 sec^{-1}), and the fluid may actually have in excess of 600 cp in the fracture (at 40 sec^{-1}). In selecting a fluid it is important to know at what shear rate the viscosity data was measured. In addition, during the testing the fluid should be sheared at a shear rate somewhat representative of the behavior expected in the fracture. This is typically on the order of 50 sec^{-1}, but for some soft rock treatments the shear rate may be much lower than this, and in some hard rock treatments, the shear rate may be much greater.

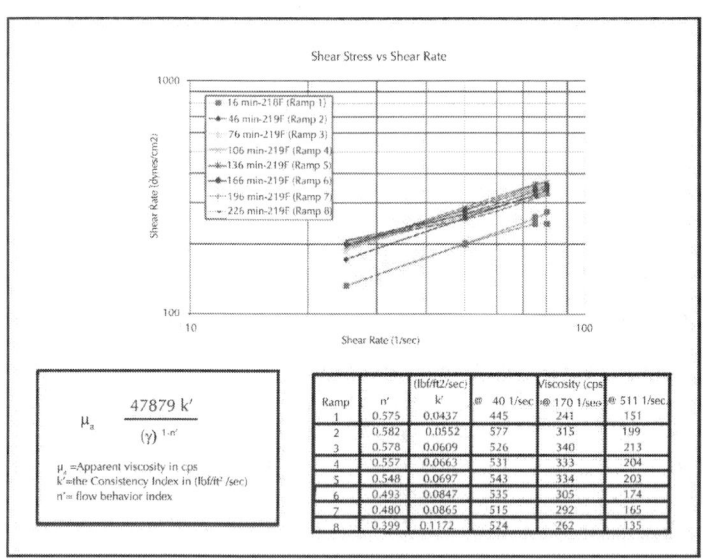

Figure 8: A set of shear stress vs shear rate set of ramps along with the calculation of apparent viscosity at three shear rates.

SHEAR HISTORY SIMULATION

As the fluid is pumped through the surface equipment, well tubular, perforations and fracture it is subjected to a range of shear rates that may have a detrimental effect on the fluid rheology. For exampleFigure 9 shows the apparent viscosity for a borate crosslinked HPG that was used to fracture a well in China. A series of premature screenouts had occurred and an evaluation was conducted to determine why. The well was completed with an open annulus and a tubing string and the treatments were being pumped down the annulus. The shear rate was calculated to be 2200 s^{-1} and the time in the tubing/casing annulus was 5 minutes. As the figure shows the apparent viscosity without the 5 minutes of high shear was 800 cps but if subjected to shear was about 20 cps. The fluid did recover its viscosity but it took 80 minutes. The higher proppant concentrations were settling out near the wellbore and causing the screenouts. The buffer package was adjusted by the service provider and that cured the problem.

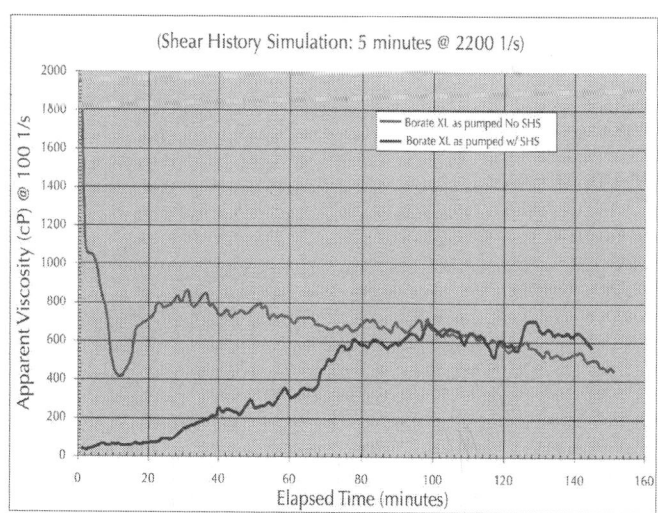

Figure 9: Viscosity Profile for a Borate Crosslinked HPG with and without shear history simulation.

Reference 13 provides a detailed procedure on how to do shear history simulation. The equipment needed is shown in Figure 10. Because the flow in the tubulars is in pipe flow rather than slot flow using a curette "Cup & Bob" viscometer at high shear rate can be misleading. The shear rate in the tubular is a function of pump rate and tubing size. The equations for determining shear rate are included in reference 13.

SLURRY VISCOSITY

Another factor affecting viscosity is the addition of proppant to the fracturing fluid to from slurry. For a Newtonian fluid the increase in viscosity due to proppant can be calculated from a equation originally developed by Albert Einstien[14]. The chart shown in Figure 11 demonstrates this effect. The figure shows that an 8 ppg slurry has an effective viscosity about 3 times that for the fracturing fluid alone. This increased viscosity will increase net treating pressure and may significantly impact treatment design. This increase in slurry viscosity also retards proppant fall as discussed below.

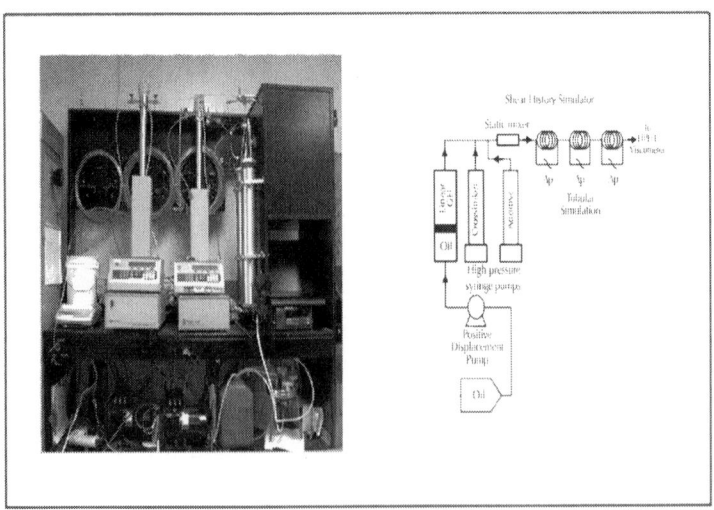

Figure 10: Shear History Simulation Laboratory Equipment.

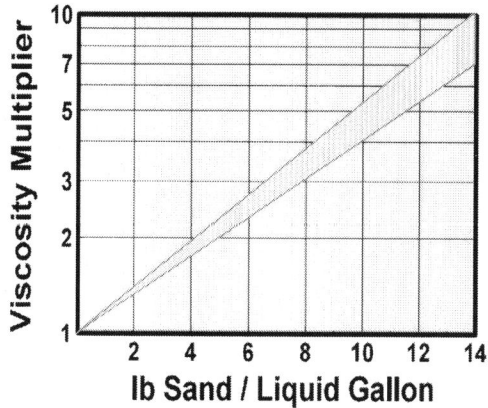

Figure 11: Slurry Viscosity Multiplier as a function of proppant concentration.

PROPPANT FALL RATES

The rate of fall for proppant is normally calculated using Stoke's Law which can be written as:

Fall Rate $= V$ (ft/sec) $= 1.66 \times 105 D2/\mu f [SGprop - SGfluid]$

Where:

D = the average proppant diameter in feet

μ_f = the apparent viscosity of the fluid in Cps

SG_{prop} = the specific gravity of the proppant (i.e. 2.65 for sand)

SG_{fluid} = the specific gravity of the fluid (i.e. 1 for water)

Stokes's Law is generally not valid for Reynolds numbers much in excess of unity[15] or for hindered settling due to proppant clustering in static fluids[16]. For crosslinked fluid the actual fall rate may be much less than Stokes Law. Hannah and Harrington[17] present lab data that shows that proppant in crosslinked fluids falls at a rate which is reduced by about 80% when compared to non-crosslinked linear gels with the same apparent viscosity. The rate of proppant fall in foams and emulsions is also much less than would be indicated by using the apparent viscosity in Stoke's Law[18]. Another factor affecting proppant fall is the particle concentration which increases slurry viscosity (Figure 11). This retards or hinders the proppant fall because of clustered settling[16] in static fluids. Finally the slurry flowing down a fracture is generally much lower that the shear rate of 170 or 511 sec^{-1} used to report the fluid apparent viscosity.

When all of these factors are put together they can significantly affect the viscosity. To provide an example consider a crosslinked gel which has a reference apparent viscosity at 170 sec^{-1} of 50 cps after four hours at reservoir temperature.

- Shear Rate Correction – If the fluid has an n' of 0.6 and the shear rate in the fracture is 50 sec^{-1}, the effective apparent viscosity in the fracture would be $(170/50)^{1-n'}$ times the measured viscosity or (1.63*50 = 81 cps).

- Slurry Correction – If the slurry enters the fracture at a concentration of 1 PPG (pounds of sand per liquid gallon) and concentrates to 10 PPG after four because of fluid loss, the average concentration of 5 PPG gives a viscosity multiple of 2 from Figure 11. This would give an effective average apparent viscosity of (2*81 = 162 cps).

- Fall Rate Correction – Harrington and Hannah[17] state that for a crosslinked fluid the rate of fall is reduced by up to 80%. For this example assume that the fall rate is reduced by 50%. This effectively doubles the viscosity to (2*162 = 324 cps).

- Temperature Correction – The fluid enters the fracture at a relatively low temperature and thus a higher viscosity. If the fluid viscosity reduces by a factor 10 over the 4 hour exposure time (down to the originally referenced 50 cps) with a log viscosity versus time relationship (typical for most crosslinked fluids) the average fluid viscosity over the four hour period would be a factor of 4.3 times the final viscosity. This gives an effective average apparent viscosity of (4.3*324 = 1393 cps).

Using a value of 1393 cps of apparent viscosity in Stoke's Law gives a total proppant fall of 15 feet during the four hour period. Almost perfect transport is achieved by a fluid system having a final reference apparent viscosity of only 50 cps.

This example may appear to be extreme but it is actually conservative. The Fall Rate Correction was reduced from 80% to 50% and the time it takes to heat up to reservoir temperature was ignored. The main point to be taken from this is that the viscosity requirements for a frac fluid can be overestimated by an order of magnitude and sufficient proppant transport can be achieved with a fluid having a reference apparent viscosity of 50 to 100 cps.

VISCOSITY AND FRACTURE TREATING PRESSURE

Treating pressure is fairly insensitive to viscosity as the pressure is proportional to viscosity raised to the ¼ power. However as discussed above the viscosity estimate can easily be off by an order of magnitude which can have a drastic impact on treatment behavior. An order of magnitude would be ($10^{0.25} = 1.8$) so the treating pressure would be 80% greater than anticipated. This could cause undesired height growth and result in treatment failure. For jobs where the control of net pressure to prevent height growth is important, fluid viscosity is a critical parameter.

REFERENCES

1.   R. P Api, M, Recommended Practice for Measurement of Viscous Properties of Completion Fluids, 01Jul-2004

2.   R. P Api, Recommended Practice for Presenting Performance Data on Cementing and Hydraulic Fracturing Equipment, 01Feb-1995

3.   GidleyJohn L., et.al., Recent Advances in Hydraulic Fracturing SPE Henry L. Doherty Series Monograph 12Chapter 7, 1555630201989

4.   ElyJohn W., Stimulation Engineering Handbook, PennWell Publishing Company, 0878144171994

5.   M. J Economides, and K. G Nolte, Reservoir Stimulation- Third Edition", John Wiley and Sons, LTD, 0-47149-192-62000

6.   M. J Economides, and T Martin, Modern Fracturing- Enhancing Natural Gas Production", ET Publishing, 978-1-60461-688-0BJ Services Company 2007

7.   KingGeorge E., "Hydraulic Fracturing 101: What every Representative, Environmentalist, Regulator, Reporter, Investor, University Researcher, Neighbor and Engineer Should Know About Estimating Frac Risk and Improving Frac Performance in Unconventional Gas and Oil Wells" SPE 152596 presented ath the 2012 Hydraulic Fracturing Technology Conference, 68February 2012

8.   KingGeorge E., "Estimating Frac Risk and Improving Frac Performance in Unconventional Gas and Oil Wells", internal Apache Corporation document, 88 23Jan 2012

9.   BeckwithRobin; "Depending on Guar- For Shale Oil and Gas Development", Journal of Petroleum Technology, December 20124455

10.   ConwayMichael W., Almond, Stephen W., Briscoe, James Earl, Harris, Lawrence E., Halliburton Services, "Chemical Model for the Rheological Behavior of Crosslinked Fluid Systems", Journal of Petroleum Technology, 352315320Feburary 1983

11.   ShahSubhash and Fagan, John; ' Fracturing Fluid Characterization Facility (FFCF): Recent Advances" DOE/MC/2907795C0490 presented at the 1995Natural Gas RD&D Contractor's Review Meeting, April 4-6, Baton Rouge, Louisiana.

12.   R. P Api, Recommended Practice on Measuring the Viscous Properties of a Cross-linked Water-based Fracturing Fluid", third edition, May 1998

13.   ISO   13503-1:2003(E)   International   Standard   "Part 1Measurement of viscous properties of completion fluids"www.iso.org,2003

14. A Einstein, Ann. Phys. 19, 289 (1906); 34, 591 (1911), b) The second order calculation is developed by G.K. Batchelor and J.T. Green, J. Fluid Mech. 56, 401 (1972

15. M. A Blot, and W. L Medlin, Theory of Sand Transport in Thin Fluids", SPE 14468 presented at the 1985SPE ATC, Las Vegas, Nevada, 2226September.

16. D. E Mcmechan, and S. N Shah, Static Proppant-Settling Characteristics of Non-Newtonian Fracturing Fluids in a Large-Scale Test Model", SPE 19735, SPE Production Engineering Journal, 63August 1991

17. HarringtonLarry, Hannah, Robert and Williams, Dennis, "Dynamic Experiments' on Proppant Settling in Crosslinked Fracturing Fluids", SPE 8342 presented at the 1979SPE ATC Las Vegas, Nevada, 2326September.

18. P. C Harris, et.al., "Measurement of Proppant Transport of Frac Fluids", SPE 95287 presented at the 2005SPE ATC Dallas, Texas, 912October.

Chapter 3

High Shear Mixers: A Review of Typical Applications and Studies on Power Draw, Flow Pattern, Energy Dissipation and Transfer Properties

Jinli Zhang, Shuangqing Xu, and Wei Li

School of Chemical Engineering and Technology, Tianjin University, Tianjin 300072, PR China

ABSTRACT

High shear mixers (HSMs), characterized by their highly localized energy dissipation, are widely used in process industries for dispersed phase size reduction and reactive mixing. Research findings on typical applications of HSMs have been summarized in this paper, namely liquid–liquid emulsification, solid–liquid suspension and chemical reactions, with an emphasis on the emulsification due to relatively intensive research in this area. The design and control of HSMs as chemical reactors need comprehensive knowledge of both the reactions kinetics and the HSMs hydrodynamics. Therefore, hydrodynamics of HSMs in terms of power draw, flow pattern and energy dissipation are then particularly reviewed from both experimental fluid dynamics (EFD) measurements and computational fluid dynamics (CFD) simulations. Limited reports on the mass and heat transfer properties in HSMs are also introduced to demonstrate their potential applicability to intensify chemical reaction processes. Due to difficulties and challenges emerged in the experimentations, CFD tools play an important role in the design, optimization and scale-up of HSMs, yet the prediction accuracies still need to be improved.

INTRODUCTION

High shear mixers (HSMs), also known as high shear reactors (HSRs), rotor–stator mixers, and high shear homogenizers, are characteristic of high rotor tip speeds (ranging from 10 to 50 m/s), very high shear rates (ranging from 20,000 to 100,000 s^{-1}), highly localized energy dissipation rates near the mixing head, and relatively higher power consumptions than conventional mechanically stirred vessels [1], which are attributed to the centrifugal forces generated from the relative motion between the rotor and the stator equipped with narrow spacing (ranging from 100 to 3000 μm). HSMs have been widely used in energy intensive processes such as homogenization, dispersion, emulsification, grinding, dissolving, and cell disruption in the fields of agricultural and food-manufacturing, and chemical reaction processes, etc. Practical applications of HSMs can be classified into three typical categories:

- Liquid–liquid emulsification, to produce less viscous liquid–liquid dispersions [2], [3], [4] and [5], highly viscous emulsions like mayonnaise [6] and bitumen [7], and miniemulsions with controlled droplet sizes for desirable polymerization [8] and [9], etc.

- Solid–liquid suspension, to produce uniform and stable nanoparticles suspensions with desirable rheology in the manufacture of healthcare, medicine, and electronic products [10], [11], [12],[13] and [14]; to produce micronized waxes dispersions used in ink, toner and coatings [15] and the pigment or dye-based solid inks [16]; to handle the wet milling of active pharmaceutical ingredients (APIs); to disrupt cells in fermentation recovery processes, and even to control the polymorph transformation during the crystallization of pharmaceutical products [17].

- Chemical reactions, to produce fine chemicals or intermediates, e.g. the patented production of chlorobenzene [18], linear alkylbenzenes [19], toluene diisocyanates [20] and [21], aniline and toluenediamine [22], and suspension polymerization to prepare small particles used in electrostatic powder coating, fluidized bed coating and plastisols [23], etc.

There are many geometric variations in the HSM designs, which can be mainly classified into in-line and batch units. The commercial in-line HSMs are usually designed as either the rotor–stator teethed or the blade-screen configuration; while the batch units have either radial-discharged or axial-discharged types (see Fig. 1). The rotor–stator assembly (also called a *generator*) can be operated in a batch, semi-batch or continuous mode. For a batch or semi-batch operation, HSMs can be used as standalone, or together with an auxiliary conventional impeller (especially for large vessels) so as to enhance the bulk mixing through creating circulation flow besides the highly localized energy dissipation generated by HSMs (Fig. 2a–c). The batch HSM can be cooperated with the in-line HSM to improve the product quality and reduce the processing time, where the in-line unit functions in a circulation loop downstream of a tank equipped with the batch unit (Fig. 2d).

Figure 1: Geometric variations of commercial high shear mixers: (a) the teethed in-line unit (Ytron-Quadro z, from Özcan-Ta kin et al. [78]); (b) the blade-screen in-line unit (Silverson 150/250 MS, from Hall et al. [4]); (c and d) the radial-discharged units (Left – Silverson L4R, from Atiemo-Obeng and Calabrese [1] and Right –VMI Rayneri Turbotest, from Doucet et al. [65]) and (e) the axial-discharged unit (Greerco 1.5 HR, from Myers et al. [64]).

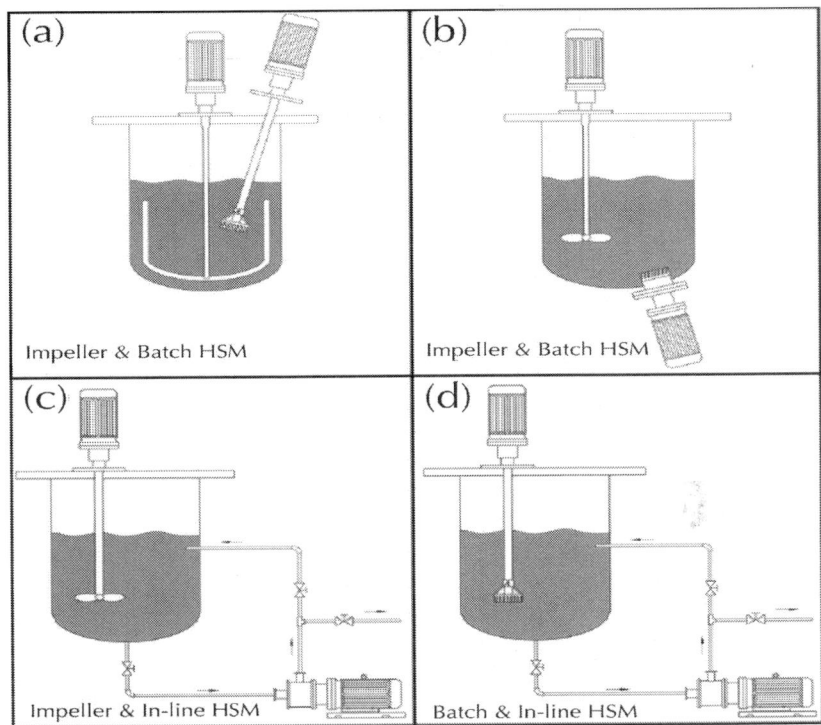

Figure 2: Feasible combinations of high shear mixers and conventional impellers.

Although the two consortia, namely the British Hydromechanics Research Group and the High Shear Mixing Research Program at the University of Maryland, have done much work about HSMs, Atiemo-Obeng and Calabrese [1] have emphasized that *"the current understanding of rotor–stator devices has almost no fundamental basis"*. Practically, the process development, scale-up, and operation of HSM are mostly relied on engineering judgments and trial-and-errors other than sound engineering principles, which lead to higher development costs, start-up problems, lost time to market and considerable material wastes[2].

In this paper, typical applications of HSMs involving the liquid–liquid emulsification, solid–liquid suspension and chemical reactions are summarized to assess the behavior performance of HSMs considering major influence factors such as operating conditions and materials properties. Then the power draw characteristics, flow pattern and energy dissipation in both batch and in-line HSMs are particularly reviewed based on both experimental fluid dynamics (EFD) measurements and computational fluid dynamics (CFD) simulations, in order to understand deeply the hydrodynamics in HSMs. Limited reports on the mass and heat transfer properties in HSMs are also introduced to demonstrate their potential applicability to intensify chemical reaction processes.

TYPICAL APPLICATIONS

Liquid–Liquid Emulsification

HSMs are widely used in cosmetics, paint, food, pharmaceutical and chemical industries to produce emulsions with narrow droplet size distributions (DSDs), as well as to create small droplets with large interfacial areas. Table 1 compares the characteristics of HSMs with those of static mixers, agitated vessels, valve homogenizers and ultrasonics, in terms of the energy dissipation range and the size distribution range, which suggests that smaller droplet size is produced as the local energy input increases. The relationship between the maximum droplet size and the local power draw is also displayed in Fig. 3[1] and [24].

Table 1: Performance features of various dispersion devices

Type of devices	Energy dissipation range (m^2/s^3)	Typical size range (μm)
Static mixers	10–1000	50–1000
Agitated vessels	0.1–100	20–500
High shear mixers	1000–100,000	0.5–100
Valve homogenizers	$\sim 10^8$	0.5–1
Ultrasonics	$\sim 10^9$	0.2–0.5

From Atiemo-Obeng and Calabrese [1].

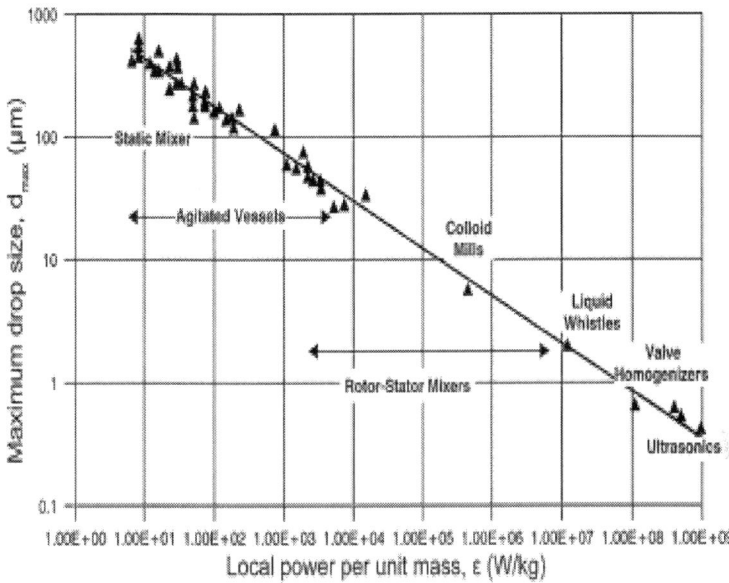

Figure 3: Maximum drop size d_{max} versus local power draw for dilute oil-in-water dispersions. The local power per mass of fluid is the total power input divided by the mass of fluid in the high intensity dispersion region of the mixer.

From Davies [24] and Atiemo-Obeng and Calabrese [1].

Owing to the variety of the HSM configurations and the liquid–liquid emulsion systems, there are plenty of investigations on measuring the droplet size ranges and their sensitivities to the operational and material parameters, discussing the dominating droplet breakup mechanisms, and developing applicable droplet size correlations for the process design and scale-up. Most of reports on liquid–liquid emulsification utilized the radial discharged batch HSMs [2], [3], [5], [6], [8], [9], [25], [26], [27], [28], [29] and [30] except for a few in-line HSMs [4], [7], [31], [32], [33] and [34], since the batch HSMs have the advantage of introducing easily the dispersed phase and monitoring the droplet size. In the case of in-line HSMs, a single pass operation mode is often utilized; and the dispersed phase can be introduced by the pipeline mixing Tee [7] and [32], direct injection into the mixing head [33], or pre-dispersion in a feed tank [4] and [34].

Calabrese and co-workers [2], [3], [25], [26], [27] and [28] carried out intensive studies on the emulsification in the absence and presence of surfactants, with different batch HSMs configurations. Initially they reported a more pronounced effect of the stator geometry on mean droplet sizes than the shear gap width in dilute low-viscosity dispersions and the measured DSDs were close to the Kolmogoroff microscale (η_K), suggesting the existence of multiple breakage mechanisms [2], [25] and [27], unlike the droplet breakage mechanisms reported in the early publications involving the turbulent fluctuating velocity [29] or the shear forces within the shear gap [5]. Then, Phongikaroon [26] extended the work by studying the effects of the viscosities of the continuous and dispersed phase as well as the interfacial tension. Padron [3] investigated the effect of surfactant on the DSDs both below and above the critical micelle concentration. It was stated that the surfactant not only lowered the interfacial tension and protected the newly formed droplets against coalescence, but also modified the interfacial rheology of the system. Phongikaroon et al. [28] conducted an application of the batch HSM to polyurethane dispersions, in which an internal emulsifier was obtained by the reaction between the pre-bounded hydrophilic group and the added base group. As many small droplets were produced, the resulting DSDs were often reported to be log–normally rather than normally distributed in volume [3], [25] and [26].

While for the single-pass in-line HSMs, Thapar [32] obtained broad DSDs due to the different turbulence and shear levels within the mixer. A critical rotational speed was found above which no significant droplet size reduction was observed. Kevala et al. [33] found that the shape of the DSD varied with the droplet viscosity, and shifted between monomodal and bimodal as increasing rotational speed and residence time. Similar to the batch HSMs [2], [25] and [27], the in-line HSMs can generate the droplet sizes lowered to the same order of η_K [4], and there exists a significant relationship between the droplet size and the stator geometry, comparing with the shear gap [32].

Emulsification process design and scale-up require the droplet size correlations with enough accuracy. The mechanistic models for droplet size correlations in stirred vessels [35], [36], [37], [38], [39],[40] and [41] and static mixers [42], [43], [44], [45], [46] and [47] can be extrapolated to HSMs, since their fundamental basis of the forces balance during droplet breakage is device independent [40].

Mechanistic understanding of the droplet breakup was often based on the pioneering work of Kolmogoroff[48] and Hinze [49], with the assumption of homogenous isotropic turbulence. Theoretically, the maximum droplet size (d_{max}) was related to the local maximum turbulence dissipation rate (d_{max}) assuming negligible droplet coalescing. For geometrically similar turbulent systems with constant Power number, the dependence on d_{max} could be rearranged into a function of the Webber number, which was an important parameter in emulsifications and suggested by Hall et al. [34] for the emulsification process scale-up in in-line HSMs. For practical applications, the Sauter mean diameter ($d_{3,2}$) was often used instead of d_{max}, due to its direct relation to the dispersed phase content and the interfacial area per unit volume. Considerable experimental evidence showed that d_{max} was proportional to $d_{3,2}$ [2], [3], [25], [26], [32], [35] and [36]. The correlations of $d_{3,2}$ are greatly dependent on the relative magnitude between the droplet size and η_K. Padron [3] summarized mechanistic models for correlating $d_{3,2}$ in different sub-ranges, i.e. the inertial sub-range ($L_T \gg d \gg \eta_K$), and the viscous sub-ranges with the dominant droplet breakup forces of the inertial stresses ($d < \eta_K$) or the viscous stresses ($d \ll \eta_K$):

$$(L_T \gg d \gg \eta_K) \quad \frac{d_{3,2}}{D} = C_1 We^{-3/5} \left[1 + C_2 Vi \left(\frac{d_{3,2}}{D} \right)^{1/3} \right]^{3/5}$$

$$(1)$$

$$(d < \eta_K) \quad \frac{d_{3,2}}{D} = C_1(WeRe)^{-1/3}\left[1 + C_2 ViRe^{1/2}\left(\frac{d_{3,2}}{D}\right)\right]^{1/3}$$

$$(2)$$

$$(d \ll \eta_K) \quad \frac{d_{3,2}}{D} = C_1 We^{-1} Re^{1/2}\left[1 + C_2 ViRe^{-1/4}\right]$$

$$(3)$$

where L_T is the turbulent macro length scale; d is the droplet size; D is the nominal rotor (impeller) diameter; We, Re and Vi are the Webber number, the Reynolds number and the viscosity group. For inviscid droplets or highly viscous droplets, these three models could be simplified with the limits of $Vi \to 0$ or $Vi \to \infty$. The effect of the dispersed phase content (Φ) can be accounted for by the modification of mechanistic models with a function of Φ, when the dispersion was well stabilized against coalescence and the rheological behavior did not alter [40].

Alternatively, $d_{3,2}$ can be empirically correlated with the energy density, which describes the effect of process parameters applied to a certain emulsion volume and accounts for the mean residence time acting on the droplet in the assumption of no re-coalescence [50].

$$d_{3,2} \propto E_V^b = \left(\frac{P}{Q}\right)^b$$

$$(4)$$

where E_V is the energy density, P is the total power input and Q is the volumetric throughput.

Table 2 lists some available theoretical, empirical or semi-empirical correlations for mean droplet diameters in HSMs, along with the features and comments from the original researches. Most models were derived from systems with negligible droplet coalescences except the one from Gingras et al. [7], which was newly developed and implicitly contained an energy density term.

Table 2: A summary of the mean droplet diameter $d_{3,2}$ correlations in high shear mixers

Mean droplet diameter correlations	HSM type	Features and comments	Reference
$$\frac{d_{3,2}}{D} = 0.038 We^{-0.6} \quad (d > \eta_K)$$ $$\frac{d_{3,2}}{D} = 0.0037\left(We^{-1}Re^{1/2}\right) \quad (d < \eta_K)$$	Batch	Dilute emulsions without surfactant were studied with low content and viscosity of the dispersed phase. DSDs were measured in situ using a video probe	[2] and [27]
$$\frac{d_{3,2}}{D} = 0.055 We^{-3/5}\left[1 + 2.06Vi\left(\frac{d_{3,2}}{D}\right)^{1/3}\right]^{3/5} \quad (d > \eta_K)$$ $$\frac{d_{3,2}}{D} = 0.093(WeRe)^{-1/3}\left[1 + 24.44ViRe^{1/2}\left(\frac{d_{3,2}}{D}\right)^{1/3}\right]^{1/3} \quad (d < \eta_K)$$	Batch	Clean and surfactant-laden systems were compared. DSDs were measured via video microscopy and image analysis. Modifications of physical properties enabled $d_{3,2}$ from both the clean and surfactant-laden systems to be correlated together	[3]
$d_{3,2} \propto (1+20\Phi)(WeRe^4)^{-1/7}$	In-line	Kerosene–water system without surfactant was confirmed to be non-coalescing even at the moderately high kerosene concentration. DSDs were measured by both the video probe and the sampling-dilution technique	[32]

$d_{3,2} \propto \left(\frac{N^3}{Q}\right)^{-0.55} \Phi^{4.13} \left(\frac{\mu_d}{\mu_c}\right)^{-0.53}$	In-line	Concentrated, coalescing and temperature sensitive bitumen emulsions were prepared in pilot scale. Cationic surfactant was added. DSDs were measured by the sampling and laser granulometer measuring. The energy density term was included in a new correlation	[7]
$\frac{d_{3,2}}{D} = 0.250(1 + 0.459\Phi)We^{-0.58}$ $\frac{d_{3,2}}{D} = 0.201\left(\frac{\mu_d}{\mu_c}\right)^{0.066} We^{-0.6}$ $d_{3,2} \propto E_V^{-0.45}, \quad d_{3,2} \propto E_V^{-0.39} \quad \text{(constant Q)}$	In-line	Emulsions were prepared in pilot scale with low to moderately high viscosity and content of the dispersed phase. Surfactant was added. DSDs were obtained by sampling and dilution technique. The empirical model of energy density was also used for correlation	[4]
$\frac{d_{3,2}}{D} = 0.29We^{-0.58} \quad (\text{high }\mu_d)$ $\frac{d_{3,2}}{D} = 0.41We^{-0.66} \quad (\text{low }\mu_d)$	In-line	HSMs in the lab and industrial scales were utilized to produce dilute emulsions of different drop viscosities in surfactant-laden systems. Constants for the Webber number model varied with the drop viscosities but were similar in both scales	[34]

For non-coalescing or slowly coalescing emulsion systems, geometric similarity was recommended in scale-up [40]. However, this is not strictly obeyed for HSMs, as the shear gap width remains the same in most cases [1]. Therefore, CFD tools are necessary for the prediction of turbulence and shear levels under different scales, and for the analysis of their sensitivities to material and operational parameters. As for coalescing systems, the extrapolation of the mechanistic models was risky and was not recommended due to the lack of fundamental basis [40]. Studies on coalescing emulsion productions in HSMs suggested the processing time be considered in the scale-up [6] and [7], which was also proposed by Leng and Calabrese [40] for the scale-up of emulsifications in stirred vessels. For practical HSM applications involving high fluid velocities, small daughter droplet sizes and high dispersed phase fractions, droplet size measurement becomes rather difficult. In these cases, it is of great help to combine CFD with the population balance equations involving prior validated droplet breakup and coalescence kernels to disclose the droplet sizes evolution throughout the mixer.

Solid–Liquid Suspension

Uniform and stable nanoparticles suspensions with desirable rheology can be obtained via HSMs. Much work in the field of aggregated-clusters breakup and suspension have been done by the British Hydromechanics Research Group and their collaborators, due to the large potential for nanomaterials to be formulated into numerous products including scratch or abrasion-resistant transparent coatings, nano-fluids, polishing slurries and environmental catalysts.

Depending on the relative magnitude of the bonds that hold primary particles together and the interactive stresses, particles' breakup occurs through three different mechanisms; i.e., (1) *erosion*, when small fragments are sheared off from large agglomerates, leading to more small particles and progressively smaller agglomerates; (2) *rupture*, when large agglomerates are broken up into smaller but fairly equisized agglomerates in a step wise process until the smallest particle size is achieved; (3) *shattering*, when large agglomerates are broken up straight away into lots of small particles but without the intermediate sizes [13]. Typical particle size distributions (PSDs) resulting from different breakup modes were schematically given by Özcan-Taşkin et al. [13], as shown in Fig. 4.

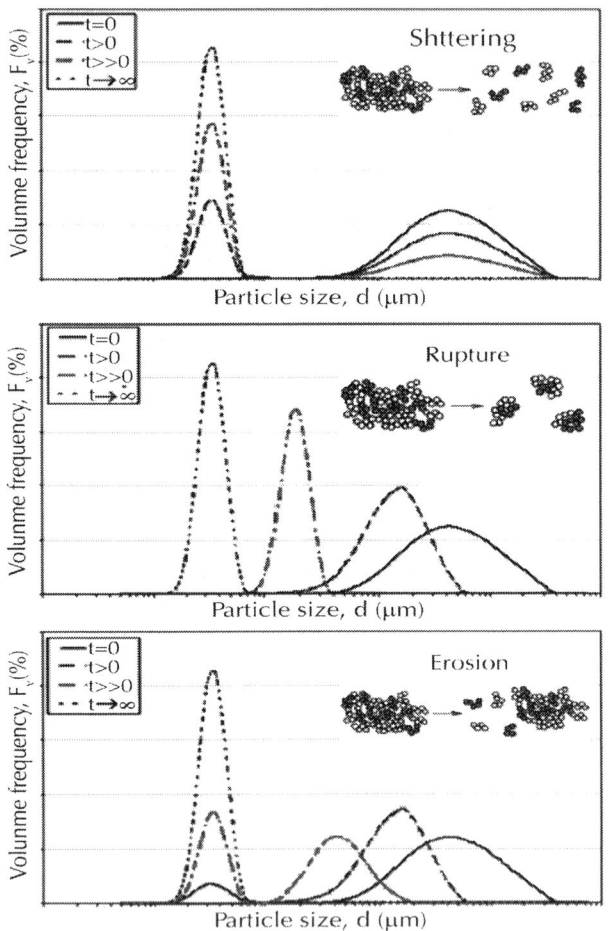

Figure 4: Schematic of particle size distributions for different breakup modes.

From **Özcan-Taş**kin et al. [13].

Aqueous suspensions of fumed silica particles, Aerosil 200 V, were obtained by breakup of large agglomerates in high shear flows by Baldyga et al. via an in-line HSM [10]. Effects of process parameters such as the suspension concentration, the rotor speed, and the number of tank turnovers on the agglomerate size were investigated. Typically bimodal PSD during the deagglomeration process indicated that the breakage of silica nanoparticle agglomerates could be described by an erosion mechanism. CFD modeling with population balance was introduced to predict the PSD, with the effects of the agglomerate structure and size distribution on suspension viscosity included. A similar study was conducted in a high-pressure nozzle disintegrator and an ultrasonic device by Baldyga et al. [11] and [12] for comparison purpose. Agglomerate breakage mechanisms taken into consideration in the simulations included: turbulent stresses in the in-line HSM; turbulent stresses and cavitation stresses in the high-pressure device; turbulent stresses, cavitation stresses and oscillations generated by ultrasonic waves in the ultrasonic device. Numerically predicted volumetric averaged particle sizes, $L_{3,0}$, were found to be 0.7–1 μm for the in-line HSM, 0.5–0.7 μm for the high-pressure system, and 0.09–0.11 μm for the ultrasonic device; but higher energy densities were required in the case of the high-pressure system and the ultrasonic device, as mentioned above in Table 1 and Fig. 3. Özcan-Ta kin and co-workers [13] and [14] examined the effect of particle type on the breakup mechanisms using an in-line HSM to process fumed silica and aluminium oxide. The authors concluded that the breakup mechanism depended on the relative magnitude of hydrodynamic stresses and material properties, specifically the tensile strength of the agglomerates. The results indicated that breakup occurred predominantly through erosion and shattering in the case of silica and aluminium oxide respectively, suggesting a prominent effect of the material properties while hydrodynamic conditions only affected the generation rate of the fine materials. Extended study in terms of HSMs configurations, operation conditions and particle types needs to be done to examine the decisive role between the hydrodynamics and material properties. Applicable PSD data correlations are also needed in order to develop, optimize and scale-up the suspension process in geometrically similar HSMs. The use of HSMs has also been on the rise in the pharmaceutical sector. Meaningful attempts by the R&D for Bristol-Myers Squibb Co. were reported by Shelley [17]. The use of HSMs in the pharmaceutical production was summarized in three main areas: to improve particle-size reduction of active pharmaceutical ingredients (APIs) through high-shear wet milling; to improve fermentation recovery processes (by using high shear to rapidly break biological cells), and to facilitate crystallization (by using high shear to create highly supersaturated starting liquids). One promisingly novel application of HSMs was reported to adjust the polymorph transformation of drugs, i.e., to produce the desirable bioavailable polymorph of a given drug compound against other polymorphs in the pharmaceutical productions. It is necessary to study further the reason that HSMs affect the polymorph transformation of drugs.

Chemical Reactions

HSMs have the advantage to intensify chemical reaction processes with fast inherent reaction rates but relatively slow mass transfer rates [51], due to their locally intense turbulence in the small shear gap with a short residence time. For heterogeneous multiphase reactions, mass and heat transfer can be facilitated as a result of the relatively large interphase areas provided by high shear mixing. Companies have patented fine chemicals and intermediates productions via HSMs, e.g. chlorobenzene [18], linear alkylbenzenes [19], toluene diisocyanates [20] and [21], aniline and toluenediamine [22], etc. Besides, nano-$CaCO_3$ [52] and $MgSO_4 \cdot 5Mg \, (OH)_2 \cdot 3H_2O$ (MOS) whisker [53] can also be prepared in HSMs.

Bourne and co-workers [54] and [55] evaluated the suitability of a commercially available HSM as a chemical reactor using the fast competitive azo-coupling reactions of 1-naphthol and diazotized sulfanilic acid. Their results indicated that, it was necessary to feed at or near the shear gap to take advantage of the high energy dissipation rate in the gap. Variations and modifications of commercial units should be made for this purpose.

Micromixing efficiency of a HSM was also investigated using a parallel competing iodide–iodate reaction system [56], [57], [58] and [59]. Characterized by the segregation index X_s (with $X_s = 0$ and $X_s = 1$ indicating perfect micromixing and total segregation, respectively), the micromixing efficiency was studied under different operating conditions such as the rotor speeds and reagent concentrations. X_s was found decreased with the increase of the rotor speed, as well as the decrease of the injected H^+ concentration. Different types of rotor–stator combinations had the similar micromixing efficiency when operated at relatively high rotor speeds (over 1200 rpm). Mixing time in the HSM was estimated to be about 10^{-5} s, shorter than that of rotating packed beds (typically about 10^{-4} s), showing an excellent micromixing efficiency of HSMs.

In the case of competitive reaction systems frequently encountered in chemical industry, it is important to have good understanding of the reaction kinetics before operating the HSMs as a chemical reactor. The reactants conversion and target product selectivity may decrease owing to the flow field characteristics of recirculation and re-entrainment in both batch and in-line HSMs and bypassing in in-line units (see Section4). Sometimes, specially designed HSMs or modification to commercially available ones will be required to ensure both mixing efficiency as well as the selectivity and overall yield of target product. It is crucial to consider the HSMs configuration design, the motor selection, the reactor control and operation, and the mechanical details when using HSMs as chemical reactors.

POWER DRAW CHARACTERISTICS

Power draw is a characteristic parameter to describe the dynamic mixing state, which is fundamental to motor selection in the mixer design, optimization and scale-up. In conventional stirred tanks, power draws are often measured by the electric, calorimetric, torque or strain method [60] and [61] and represented by a dimensionless Power number Po. In the laminar regime, Po is inversely proportional to Reynolds number Re. In the turbulent regime, Po is often constant and varies from about 1 to 6. The transition from laminar to turbulent flow occurs around $Re \sim 10^4$ [62].

Power draws in HSMs vary with the HSM geometry, the size of the rotor, the rotational speed, and properties of the process fluids. In the case of batch HSMs, power draws are also functions of the vessel size, the internal structures (with or without baffles, obstacles, etc.), and the mounting positions (centered or off-centered).

Batch HSMs: Power Draw Correlations and General Power Curves

For batch HSM systems (mainly radial discharge or axial discharge impellers), the power draw can be measured in a similar way as stirred vessels. Unlike the primary length scale of the impeller (rotor) diameter to define Re and Po in conventional stirred tanks, there are several characteristic lengths needed considering in HSMs, including the nominal rotor diameter D, the clearance between the rotor and stator (or the shear gap width) , and the hydraulic radius of the stator slots/holes R_h [1] and [63].

Table 3 summarizes the dimensionless Power number correlations for Newtonian fluids in batch HSMs. In the light of the conventional definition of Re and Po based on the nominal rotor diameter D (except in the last reference listed in Table 3), the power curves for Newtonian fluids in batch HSMs are similar to that of stirred vessels, displaying about the same range of Po [1]. Nevertheless, the actual power draws in HSMs are higher since they are usually operated at increased rotational speeds.

Table 3: A summary of Power number correlations for Newtonian fluids in batch high shear mixers

HSM configurations	Power number correlations	Reference
Greerco 1.5 HR	$Po = 700/Re$, $Re < 100$; $Po \equiv 1.4$–2.3 in turbulent regime	[64]
Ross ME 100LC and Silverson L4R	$Po \propto 1/Re$, independent of stator geometry in laminar regime; $Po \equiv 2.4$–3.0 for the Ross mixing head; $Po \equiv 1.7$–2.3 for the Silverson mixing head in turbulent regime	[63]
VMI Rayneri with straight blades	Rotor–stator: $Po = 314Re^{-0.985}$, $Re < 100$; $Po \equiv 3$ in turbulent regime; Rotor only: $Po = 92.7Re^{-0.9983}$, $Re < 100$; $Po \equiv 3$ in turbulent regime	[65]
Dual shaft (Paravisc impeller& VMI Rayneri with curved blades[a])	HSM only: $Po = 138/Re$, $Re < 100$; Paravisc only or Dual shaft: $Po = 368/Re$, $Re < 100$	[66]

[a]When considering the HSM, Re and Po definitions were based on the outer diameter of the stator, rather than the nominal rotor diameter used in other references listed here.

Myers et al. [64] measured the power draw of an axial-discharged HSM (Greerco 1.5 HR, $D = 84$ mm, $\delta = 0.25$ mm) in the laminar, transition and turbulent regime. Experiments were conducted in a flat-bottom vessel with the HSM centrally-mounted. Newtonian aqueous solutions of glycerine and corn syrup, and non-Newtonian shear-thinning aqueous solutions of carboxymethylcellulose were used as working fluids. Geometry effects on turbulent power draw were firstly investigated using water as the working fluid. It is indicated that in turbulent operation, the power draw was influenced primarily by the pumping mode (up- or down-pumping), and to a less extent by vessel baffling and other geometric parameters (off-bottom clearance, location of the upper deflector, etc.).

Padron [63] acquired power draw data for two radial-discharged, bench scale HSMs (Ross ME 100 LC, D = 34 mm, δ = 0.5 mm; and Silverson L4R, D = 28 mm, δ = 0.2 mm) off-centered in an unbaffled tank with Newtonian working fluid. Power number was found inversely proportional to Re and somewhat independent of stator geometry in the laminar regime. The Silverson mixer drew slightly higher power at constant Re due in part to its smaller gap width [1]. Fully turbulent occurred above $Re \sim 10^4$, but the Ross mixer had a smaller transition region. The constant turbulent Power number was dependent on stator geometry and varied from 2.4 to 3.0 for the Ross mixing head and 1.7–2.3 for the Silverson head. In general, for a given geometry, the Power number increased with the number of openings in the stator, indicating that the Power number per stator slot was the same [1] and [27]. Padron [63] also suggested that, Po was controlled by viscous dissipation in the shear gap in the laminar regime, and by fluid impingement on the stator slot surface or turbulence in the jets emanating from the stator slots in the turbulent regime.

Doucet et al. [65] investigated power draw characteristics for a radial-discharged type HSM (VMI Rayneri Turbotest with straight blades, D = 85 mm, = 1.5 mm) centrally mounted in an unbaffled tank. Glucose solutions were used as Newtonian working fluids, while xanthan gum and sodium carboxymethyl cellulose were used as non-Newtonian working fluids. Results showed that the stator head had a significant influence on the power draw in the laminar regime, as the power constant $Kp = Po \cdot Re$ turned to be almost three times lower (KP = 92.7) when operated the mixer without the stator, compared with KP = 314 for a complete rotor–stator assembly. On the other hand, the turbulent Power number remained the same for both configurations at $Po = 3$, which were in the same range as those reported by Padron [63] in the case of a Ross HSM with a slotted stator head. These results were in agreement with the conclusions given by Calabrese et al. [1] and [27] who postulated that the power draw in the laminar regime was mainly due to the stator head, and in the turbulent regime was due to the jet discharged by the rotor.

Power draws of a dual shaft (an axially mounted Paravisc impeller, and an off-centered VMI Rayneri HSM with curved blades, $D_{OS} = 90$ mm, $\delta = 2$ mm), dished bottom stirred vessel were carried out in the laminar regime by Khopkar et al. [66] with the Newtonian fluid of glucose aqueous solution and the non-Newtonian fluid of carboxy methyl cellulose. Power draw measurements of HSM when the Paravisc impeller kept stationary were first carried out using the Newtonian fluid. With the Re and Po definition based on the outer diameter of the stator (90 mm, compared to a nominal rotor diameter of 85 mm), the power constant KP was found to be 138 in the laminar regime, i.e., approximately 2.5 times less than that of Doucet et al. [65] (still about 2.1 times less when Re and Po defined with the nominal rotor diameter), indicating higher energy efficiency of the HSM with curved blades than the one with flat blades. Power draw measurements for the Paravisc impeller with the Newtonian fluid in the case of fixed HSM and rotating HSM were determined respectively. KP of the Paravisc impeller was found to be 368 in the laminar regime. It was reported that the presence of HSM did not influence the power characteristics of the Paravisc impeller for both Newtonian and non-Newtonian fluids, indicating weak pumping capacity of HSM. However, the HSM not only initiated the mixing early but also improved the mixing rate, since the presence of eccentric impeller not only broke the symmetry of the system but also played the role of a baffle [67].

Attempts have been made to correlate power draw data from both Newtonian and non-Newtonian fluids in batch HSMs in the same power curve (termed as *General Power Curve* in this article) in order to get a convenient insight for scale-up problems [64], [65] and [66]. The apparent viscosities for Re calculation were functions of average shear rates for different kinds of non-Newtonian fluids, i.e., $\mu_a = f(\gamma_{av})$. Different methods were reported for the determination of the average shear rates, including the Metzner-Otto approach [68] (see Eq. (5)) which assumed the shear rates directly proportional to the rotational speed; the Rieger-Novák approach [69] (see Eq. (6)) which considered the shear rates as equal to the rotational speed; and the macroscopic approach for power consumptions [65] (see Eq. (7)).

$$\gamma av = KN \tag{5}$$

$$\gamma av = N \tag{6}$$

$$P = \mu_a \gamma_{av}^2 V \tag{7}$$

where N, P, V are the rotational speed, power input and volume of the mixer; γ_{av} and μ_a represent the average shear rate and apparent viscosity of the non-Newtonian fluids; K is the Metzner-Otto constant.

Although established in the laminar regime, the Metzner-Otto approach also found its applications in the transition and turbulent regime, and was considered as the most promising method to provide a good shift of the non-Newtonian power curves on the Newtonian ones with appropriately determined K values, i.e., to produce smooth and continuous *General Power Curves* [64], [65] and [66]. For conventional impellers, Nagata [70] reported that K increased from about 10 for open-style impellers (with an impeller to vessel diameter ratio $D/T \leq 0.60$) to about 25–30 for close-clearance impellers ($0.90 \leq D/T \leq 0.99$). Except for the impeller type, K also varied with the types and power law indexes of the process fluids, and even was a weak function of the rotational speed [65], [66], [71] and [72].

Operating in the laminar regime, Khopkar et al. [66] derived K values for a batch HSM from 8 to 12 with respect to the power law indexes of the fluids. Doucet et al. [65] reported a wide range of K values from 0.6 to 70 depending on the types of fluids and rotational speeds, when processing non-Newtonian fluids using a batch HSM from the laminar to turbulent regime. Myers et al. [64] got a similarly high K value of 71 to correlate well the Froude number-adjusted *General Power Curves* [73] from the laminar to turbulent regime. The particularly high impeller (rotor) to vessel (stator) diameter ratio [64] and the extended use of the Metzner-Otto equation in the transition and turbulent regime [65] were considered responsible for the unrepresentative K values as compared to that from Nagata [70].

As HSMs are commonly used to handle non-Newtonian industrial fluids, it is fundamental to determine the *General Power Curves* for different process fluids and mixer configurations. The few data sets currently available for batch HSMs are limited to devices with a single rotor with blades (not teeth) surrounded by a single stator [1]. Much work needs to be done to get a comprehensive understanding of the power draw characteristics of other typical HSMs configurations. Although the Metzner-Otto approach has been broadly used in the power draw calculations of non-Newtonian fluids, yet the Metzner-Otto constant K cannot be predicted *a priori* and a unique value cannot be used due to the large variability observed [65],[66], [71] and [72]. Complete characterization of non-Newtonian power draws require more extensive data and characterization of the liquid rheological properties at the effective shear rates that occur in the specific high shear mixing head [64]. In the case that K is found to be dependent on process parameters, a compromise might be to divide the operation parameters into intervals with a specifically determined average K value for each interval.

In-Line HSMs: Power Draw Model and Determination of Model Constants

Kowalski and co-workers [74], [75] and [76] have made detailed investigations about the power draw for in-line HSMs (operated in the continuous mode). In-line HSMs differ from batch ones because the flowrate is usually controlled independently of the rotor speed [74]. The relationship between the rotor speed, the flowrate and the energy dissipated in in-line HSMs cannot be adequately described by a single Power number. A power draw model for in-line HSMs was given by [75]:

$$P=PT+PF+PL=Po z\rho N^3 D^5+k_1 MN^2 D^2+PL \tag{8}$$

where P is the total power input into the in-line HSM related to the material properties and operating conditions; P_T the power required to rotate the rotor against the liquid in the gap; P_F the additional power requirements from the flow of liquid through the gap; and P_L the power losses by vibration and noise, kinetic energy losses at the entrance and exit, and the accuracy of measurements, etc.

Cooke et al. [74] presented a simple method for determining the constants in the power draw model shown in Eq. (8). An assumption of $P_L = 0$ was made when any systematic measurement errors and power draw losses were accounted for. When the backpressure valve was fully shut and the flow was zero, Po_z could be directly obtained by torque measurement. When the valve on the outlet was fully open, the HSM was free to pump and a maximum power draw gave the characteristic Power number Po_u under the assumption that flowrate was proportional to the swept volume of the rotor. The constant k_1 in Eq. (8) could be calculated with Po_z and Po_u.

Although this approach appeared attractive, Kowalski et al. [76] pointed out that it had two major flaws: firstly, k_1 showed considerable variation although Po_z and Po_u showed reasonable consistency; secondly, as the power draw at low (including zero) flowrate showed an anomalous increase the value of Po obtained was too large.

Kowalski [75] validated the model with data from a Siefer Trigonal mill ($D = 220$ mm, $ = 0.1–0.4$ mm) processing two slurries, namely a 60 wt% calcite slurry in water and a 50 wt% soda ash slurry in a nonionic surfactant. The power input P was calculated using the electrical method:

$$P = \sqrt{3}UI\eta \cos\varphi$$

(9)

where U, I, η, $\cos\varphi$ are the apparent voltage, the apparent current, the motor efficiency and the power factor, respectively. Data fit revealed similar values of the constants k_1 and P_L for the two slurries indicating that they were characteristics of the HSM rather than the materials. Assuming a laminar flow in the mill, $K_p = Poz \cdot Re$ were found slightly different for the two slurries. A K_p value of the order 0.1 was unexpectedly low while typical values were of the order 10–100, or even close to 1000 [64],[65] and [66].

The quality of the data fit was hampered by poorly described slurries viscosity, as well as a reliance on the electrical measurement with the constant η and $\cos\varphi$, which were actually not the case [76]. Furthermore, P_L in the power draw model varied with load and could not be regarded as a constant.

By both torque and calorimetric measurements, Kowalski et al. [76] then studied the power draw of an in-line HSM (Silverson 150/250 MS, D_{in} = 38.1 mm, D_{out} = 63.5 mm, = 0.229 mm) over a wide range of water flowrates (0–6,000 kg/h) and rotor speeds (0–11,000 rpm). Due to the advantages and disadvantages of these two methodologies [61], a better accuracy of the torque method was found at higher flowrates; whereas the calorimetric technique was more accurate at low flowrates when the temperature rise was larger. The accuracy of both measurement techniques improved with the rotor speed, therefore the ability to overspeed the modified Silverson was believed to be crucial.

Both torque and calorimetric based data showed similar trends: the power draw increased with the rotor speed and flowrate except an anomalous increase at low flowrates. A power law fit of torque based power draws versus rotor speeds yield an index of ~2.5 (see Fig. 5), which was consistent with earlier findings[55]. The authors attributed this to the interrelationship between the convection of energy from the mixer ($\propto N^2 D^2$) and the power consumption under turbulent conditions ($\propto N^3 D^5$).

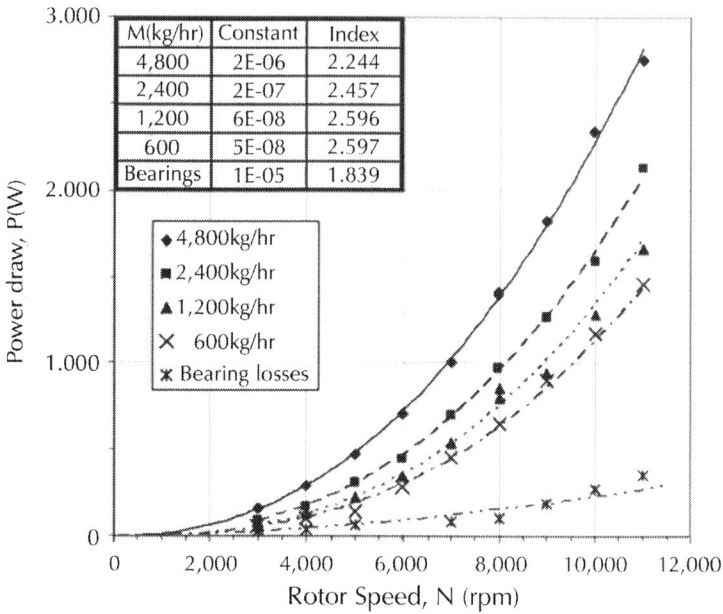

M(kg/hr)	Constant	Index
4,800	2E-06	2.244
2,400	2E-07	2.457
1,200	6E-08	2.596
600	5E-08	2.597
Bearings	1E-05	1.839

Figure 5: Power draw and bearing losses measured by the torque method as a function of the rotor speed for different flowrates.

From Kowalski et al. [76].

After the corrections on torque (bending of the shaft, resistance in the bearings subtracted) and temperature rise (temperature rise in the stationary HSM subtracted) in the experimentation, it was assumed that P_L in Eq. (8) was negligible, and the constants Po_z and k_1 were obtained by the regression and graphical methods. For both the torque and calorimetric techniques, the regression routine gave slightly lower values of Po_z but higher values of k_1. In general, Po_z ranged from about 0.197 to 0.229, and k_1 ranged from about 7.46 to 9.35.

With an analogy to a centrifugal pump, the conversion efficiency of the motor power into hydraulic power (termed as *pump efficiency*) of in-line HSMs could be represented as a function of the flowrate:

$$E_p = \frac{M}{\rho} \cdot \frac{(p_2 - p_1)}{P}$$

(10)

where E_p, M, P are the pump efficiency, the mass flowrate and the measured total power input of the in-line HSM, p_1 and p_2 are the measured pressures of the in-line HSM inlet and outlet, and is the density of the process fluid. Data fitting method gave a logarithmic linear relationship between the efficiency and flowrate (see Fig. 6). The sharply decreased pump efficiency at low flowrate was considered as one possible reason for the unexpected power draw increase. Alternatively possible cause of the anomalous power draw at low flowrates might be recirculating flows in the vicinity of the screen (stator), as will be discussed in Section 4.

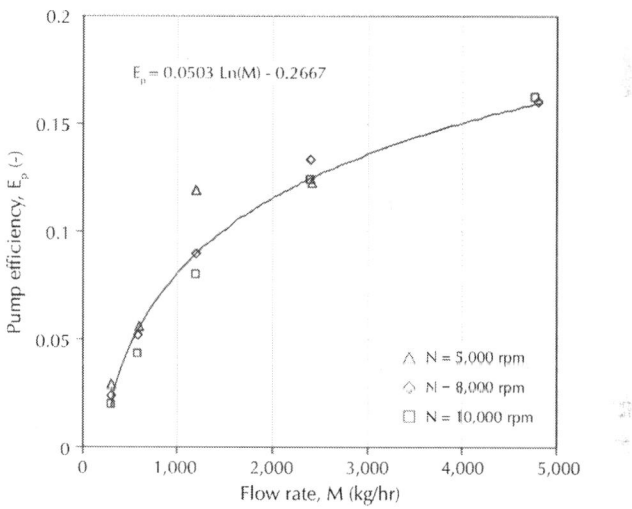

Figure 6: Pump efficiency of Silverson at different flowrates.

From Kowalski et al. [76].

Although the efficiency term was obtained by using the total measured power, it is the case when the efficiency was used to modify the "flow term" (second term in Eq. (8)) that seemed to more closely follow the measured data points. Within the flowrates ranging from about 300 to 4,800 kg/h, the pump efficiency was found no more than 0.2, indicating a low pumping capacity of the Silverson HSM. Similar comments have been made by Atiemo-Obeng and Calabrese [1]:

"Units that have a rotor with blades, such as Silverson in-line series, can simultaneously pump and emulsify/disperse material (to some extend). However, many designs, such as those with multiple rows of rotor and stator teeth, have marginal pumping capacity and may even cause a pressure drop....It is often necessary to feed the mixer with a separate pump".

Cooke et al. [77] advanced their work to investigate the power draw characteristics of an in-line HSM involving three kinds of stator arrangements (namely no screens, screens with standard holes and fine holes), with both Newtonian and non-Newtonian working fluids, and from the laminar, transition to turbulent flow regime. The Metzner-Otto approach and the experimentally determined Metzner-Otto constant enabled the prediction of the Power numbers for both kinds of fluids (see Fig. 7). In order to produce the continuous power curves from the laminar, transition to turbulent flow regime, the "power term" (first term in Eq. (8)) in the transition regime was assumed to consist of contributions by a laminar and a turbulent part, with the laminar Poz inversely proportional to Re while the turbulent Poz remained almost constant. Models constants from Eq. (8) obtained for various HSM arrangements showed fairly obvious differences, which reflected a significant influence of the stator geometry on the power draw, similar to the results reported by Pardon [63] for batch units.

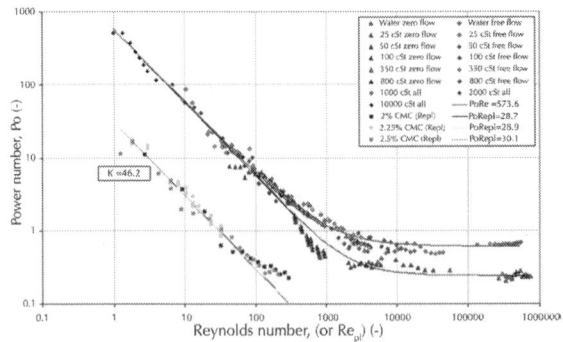

Figure 7: Power draw data for the standard Silverson 150/250 MS processing both Newtonian and non-Newtonian fluids.

From Cooke et al. [77].

Hall et al. [34] studied power draws in two different scale in-line Silverson units by the calorimetry technique when processing dilute emulsions. Their model constants Poz in Eq. (8) showed very similar values at each scale while k_1 showed quite large variation. Özcan-Ta kin et al. [78] investigated the power characteristics of three different commercial in-line HSM units using the calorimetry method. The model constants Poz and k_1 in Eq. (8) obtained for different HSM configurations were found of the same typical order of magnitude and only slight differences were observed among each other.

Due to the poorly obeyed geometric similarity and inherent inaccuracy of the calorimetry technique, the results from Hall et al. [34] and Özcan-Taşkin et al. [78] are not comparable with that from Cooke et al. [77]. The power draw model constants reported for in-line HSMs with different configurations are summarized in Table 4.

Table 4: Power draw model constants for in-line high shear mixers with different configurations

In-line HSM con-figurations	Working fluids	Measurement technique	Operating condi-tions	Po_z	k_1	References
Silverson 150/250 MS with standard screen dual emulsifier	Water	Torque and calorimetry	$N =$ 3000–11000 rpm, $M =$ 0–6000 kg/h	0.197/0.229[a]	9.35/7.46	[76]
Silverson 150/250 MS with fine screen dual emul-sifier	Water, silicon oils and CMC aqueous solutions	Torque and calorimetry	$N =$ 3000–8000 rpm, $M =$ 0–9000 kg/h	0.145	8.79	[74] and [77]
Silverson 150/250 MS with standard screen dual emulsifier			$N =$ 3000–7000 rpm, $M =$ 0–9000 kg/h	0.241[b]	7.75	

Silverson 150/250 MS with no screens			$N = 3000–6000$ rpm, $M = 0–9000$ kg/h	0.203	6.93	
Silverson 150/250 MS with GPDH-SQHS (coarse screen)	Distilled water	Calorimetry	$N = 3000–9000$ rpm, $M = 1078–5390$ kg/h	0.13	9.1	[78]
Silverson 150/250 MS with standard screen dual emulsifier			$N = 3000–9000$ rpm, $M = 1078–5390$ kg/h	0.11	10.5	
Ytron Z unit (teethed HSM)			$N = 3000–7950$ rpm, $M = 575–1800$ kg/h	0.18	10.6	
Silverson 150/250 MS with standard screen dual emulsifier	Silicon oil in water emulsions	Calorimetry	$N = 3000–11000$ rpm, $M = 50–6150$ kg/h	0.229	7.46	[34]
Silverson 088/150 UHS with standard screen dual emulsifier			$N = 4000–10000$ rpm, $M = 50–1450$ kg/h	0.254	9.59	

[a]The constants from the torque method were showed before the virgule, while those from the calorimetry method after the virgule.

[b]Power draw model constants were solely derived from the torque method although both two methods were adopted by the original authors.

FLOW FUNDAMENTALS FROM EXPERIMENTAL AND COMPUTATIONAL FLUID DYNAMICS

Controversial opinions on the criteria for HSMs scale-up and design were reported in the open publications. Vendors often design and scale-up HSMs based on constant tip speed (ND) and constant shear gap width, making the criterion equivalent to constant nominal shear rate in the shear gap [1]. Bourne and Studer [55] proposed that the scale-up procedure on the basis of constant tip speed (ND) was better than that on the basis of constant power density. On the contrary, Utomo et al. [79] argued that in the turbulent regime, scale-up of HSMs should be based on constant energy dissipation rate per unit mass (N^3D^2) and geometry similarities, whereas the constant tip speed criterion leads to lower energy dissipation rate per unit mass in the large scale.

An efficient reactor design and scale-up criterion should be capable of maintaining the desirable flow pattern and distribution of energy dissipation. Flow fundamentals involving these two aspects can be obtained by advanced non-invasive experimental fluid dynamics (EFD) techniques including laser Doppler anemometry (LDA) and particle image velocimetry (PIV). For in-line HSMs, specially manufactured units with transparent front and/or circumference of the volutes are required to allow laser beam access, which may be relatively easy to satisfy for batch systems. Geometric complexity brings difficulty in fully three-dimensional (3D) flow field measurement in the vicinity of the rotor, e.g., the commonly small shear gap width, the small slots or holes in the rotor and stator, and the relative motion between the rotor and stator. To our knowledge, rare PIV study of the flow field in HSMs was reported. Meanwhile, most published LDA studies in HSMs also only focused on the jet emanating from the stator slots/holes and the bulk flow, and only two velocity components were measured. Reported laser-based EFD measurements of batch HSMs are predominantly limited to radial-discharged types, with less attention paid to axial-discharged ones; while for in-line HSMs, only lab scale units or simplified prototype of commercial units are studied.

Numerical simulations based on computational fluid dynamics (CFD) can give the flow field at a point-to-point resolution. CFD methodology is less time-consuming when considering the geometric variations of HSMs. Compared with the multiple reference frame model (MRF) which is actually a steady-state approximation, sliding mesh method (SM) is more suitable to simulate the transients that may occur in strong rotor–stator interactions [80]. Besides, the accuracy of the flow field predictions depends on the turbulence model and near-wall treatment. As far as we know, CFD modeling of HSMs reported in the open literature so far are mainly based on the standard k–turbulence model. Although the standard k–turbulence model is robust and computationally economic [81], it has the inherent weakness such as the unrealistic assumption of isotropic turbulence, the unsatisfactory performance in the near-wall region, insensitivity to streamline curvature and rotation as well as an excessive damping of turbulence [80], [82],[83], [84], [85] and [86]. Furthermore, it has been reported that the tendency to underpredict turbulent kinetic energy is typical for all Reynolds-averaged Navier-Stokes (RANS) based turbulence models [86]. Large eddy simulation (LES) can give better prediction of turbulent kinetic energy, but it requires more computational cost and more refined computational cells [85] and [86].

In-Line HSMs: Flow Pattern and Energy Dissipation

Calabrese and co-workers [27] and [87] studied the flow field in a simplified prototype of an in-line IKA HSM containing a single set of rotor and stator teeth by LDA measurement and CFD modeling (shown in Fig. 8). The HSM had only 12 rotor teeth and 14 stator teeth, and the front of the volute was made of Plexiglas for laser beam access. Radial and tangential velocity components were acquired in the stator slots and part of the volute by two-dimensional (2D) LDA with water as the working fluid. Both angularly averaged and angularly resolved velocity fields were measured. 2D sliding mesh simulations based on the standard k–ε turbulence model together with standard wall functions were performed. However, LDA results indicated that the flow field was highly three-dimensional. Bypassing existed in the in-line HSM as a result of a clearance between the top of the rotor teeth and the front volute cover (0.2 mm compared to a shear gap width of 0.5 mm).

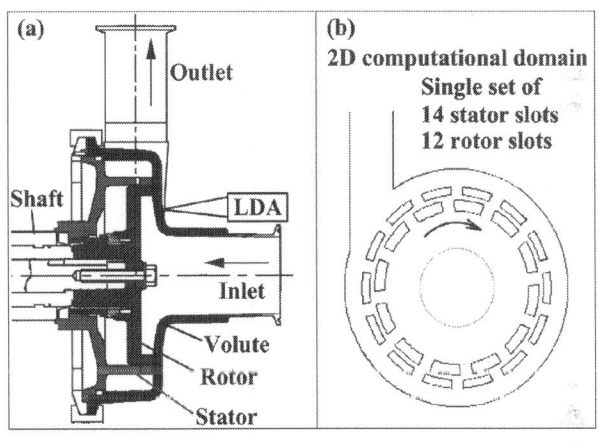

Figure 8: Schematic diagram and the 2D computational domain of the prototype IKA mixer with a single set of 12 rotor and 14 stator teeth.

From Atiemo-Obeng and Calabrese [1].

Both LDA measurements and CFD simulations indicated an extremely complex flow pattern in the HSM, with the jet emanating from the stator slots into the volute along the surfaces of the downstream stator teeth, recirculation in the stator slots and re-entrainment of the volute fluid back into the shear gap. The devices did not function well as a pump since the tangential motion was broken. Compared with the 2D simulation, LDA measurements showed stronger jet, more focused recirculation zones and stronger re-entrainment.

Turbulent kinetic energy distribution by both LDA and CFD showed that much of the fluid leaving the stator slots did not experience intense deformation field, though results from the two methods were somewhat different. It is because this bypassing that devices with multiple rows of rotor and stator teeth were more effective for dispersion. For turbulent flow through rotor–stator teethed HSM, Calabrese et al. [27]proposed that, the flow stagnation on the leading edge of the downstream stator teeth provided a major energy field for emulsification and dispersion. Further experimental and computational study of a similar device with an enlarged shear gap width showed lower level of turbulent kinetic energy. As a result, it is suggested that a narrow shear gap was favored even though the shear in the gap was not a major contributor to the dispersion process.

Discrepancies between the experimentally and numerically determined velocity fields and turbulent kinetic energy distributions from Calabrese et al. [27] and [87] can be attributed to the inappropriate 2D computational domain, the standard k–ε turbulence model and the standard wall functions, which may not be sufficient to predict the highly three-dimensional, highly rotating and wall-bounded flow in HSMs. Fully 3D sliding mesh large eddy simulation with proper near-wall treatment could be adopted for more accurate predictions.

Özcan-Ta kin et al. [78] studied the flow characteristics of three in-line HSMs configurations by 3D CFD simulations using the realizable k–ε turbulence model coupled with a steady state MRF technique. No experimental results were provided for comparisons, but the CFD prediction capabilities were expected to be limited by the RANS approach and the MRF model.

Batch HSMs: Effect of Stator Geometries

In combination with both EFD and CFD techniques, Pacek and co-workers [79], [88], [89], [90] and [91] studied the turbulent flow field of the radial-discharged batch HSM (Silverson L4R) in an unbaffled, flat-bottom vessel. A 2D LDA system was operated in back scattered mode to measure the axial and radial velocity components with silver coated hollow glass sphere seeded in water. 3D sliding mesh simulations using the standard k–ε turbulence model were employed to predict the velocity and energy dissipation rate distribution. Enhanced wall function which could describe the flow in the viscous sublayer, buffer region and fully turbulent outer region of the boundary layer was applied at the wall boundary.

Pacek et al. [88] studied the flow field in the Silverson L4R HSM fitted with a disintegrating head. CFD predicted velocity vectors also showed the presence of high velocity jets and recirculation loops induced by those jets, very similar to that reported by Calabrese et al. [27]. Flow pattern around the stator hole was affected by the position of the blade and the maximum velocity of the jet occurred when the blade was approaching the leading edge of the stator. Normalized energy dissipation rate was found independent of rotor speed and the highest dissipation rate occurred in the jet close to the leading edge of the hole when the rotor blade was approaching this edge. When the rotor blade overlapped with the leading edge, energy dissipation rate decreased but remained much higher than that in other point in the mixing head. Energy dissipation rate in the gap was smaller than in the impinging jets but lager than in the rotor swept region.

Although LDA measurements revealed similar flow pattern as CFD prediction, obvious differences existed, possibly due to the limitation of the standard k–ε turbulence model. The maximum jet velocity was underpredicted by 25%. There were also differences between the predicted and measured axial and radial velocities both near the mixing head and in the bulk region. Flow number calculated from LDA data was 0.217, higher than 0.176 from CFD prediction. Energy balance based on LDA data indicated that approximately 70% of energy supplied by rotor was dissipated around the mixing head, whereas the standard k–ε model predicted almost 80%.

Bulk mixing in the Silverson L4R HSM was found not very intensive, as velocities in the bulk were a few orders of magnitude lower than that in the jets. The calculated pumping efficiency of the

HSM [92] was about two orders of magnitude lower than that of the Rushton turbine [93], which again showed a poor bulk mixing.

Utomo et al. [79] attempted to improve the CFD prediction accuracy by refining the computational cells in the stator holes and utilizing a smaller transient iteration time step size. Though the predicted turbulent (Reranging from 26,000 to 52,000) Power number of 1.55–1.57 for the batch Silverson L4R was close to the experimental value reported by Padron [63], the modified model practically gave no improvement compared to that of Pacek et al. [88], in terms of the tangential and radial velocities, Flow number, and energy dissipation rate near the mixing head.

Utomo et al. [89] numerically investigated the effect of the stator geometry (disintegrating head, slotted head and square hole head considered) on the flow pattern with previously reported model settings. In all cases regardless of the shape and size of the stator holes, the jets only emerged from a part of the hole close to the leading edge, and behind the jets the circulation loops were formed. However, the characteristics of the jets emerging from the wide and narrow holes were different. Jets emerging from narrow holes had positive tangential velocity (opposite direction to rotation) and those from wide holes had negative tangential velocity (same direction as rotation) (see Fig. 9). The authors explained this as an interaction between the jets and circulation flow behind them.

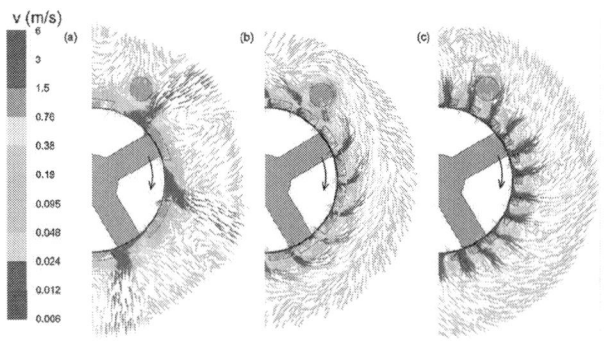

Figure 9: Predicted 2D velocity vectors (radial and tangential velocities) around the mixing heads (r = 14.3–30 mm): (a) disintegrating head, (b)

slotted head and (c) square hole head (From Utomo et al. [89]). The velocity vectors for disintegrating, slotted and square heads were taken at plane $z = -0.8$ mm, $z = 0$, and $z = 1.3$ mm, respectively.

For all geometries, the energy dissipated in the rotor swept region (about 50–60% of the total energy input) and in the jet region (about 25% of the total energy input) depended linearly on the flowrate, whereas the energy dissipated in the holes (only small fraction of total energy input in spite of high energy dissipation rate) correlated better with the total surface area of the leading and trailing edges where the stagnation occurred.

Selection principles of the stator in HSMs were suggested based on their CFD results: stators with wide holes were suitable when bulk mixing was required, as jets emerging from wide stator holes extended to the bulk region and considerable amount of energy was dissipated thereby; while stators with narrow holes could be used to produce dispersions with narrow droplet size distributions, as narrow stator holes created more uniform and concentrated energy dissipation rate in the hole region.

Underprediction of Power numbers Po was also reported (see Table 5), as compared with that from Padron[63]. For disintegrating and square hole heads, the predicted Po were approximately 10% lower than experimental values, while for slotted head the difference was up to 20%.

Table 5: Power numbers for various stator geometries at 4000 rpm

Stator geometries	Total hole area (mm²)	Po (simulation)	Po (from Padron [63])	%Difference
Disintegrating head	301.44	1.53	1.7	−10.0%
Slotted head	276.80	1.66	2.1	−20.9%
Square hole head	574.08	2.05	2.3	−10.9%

From Utomo et al. [89].

Although Utomo et al. [89] pointed out that, insufficient computational cells to account for large velocity gradients, impractically all identical stator holes in computational grids and the discretization scheme affected their simulation results, yet the prediction accuracy was limited by the standard k–$\square$ turbulence model.

Barailler et al. [94] investigated the hydrodynamics of a batch HSM VMI Rayneri Turbotest in the laminar regime with viscous Newtonian fluids, based alternatively on the virtual finite element method (VFEM). Numerical simulations revealed similar flow pattern found in other radial-discharged HSMs, namely a discharge of the fluid through the stator at the front of the blade and an intake of fluid behind the blade (shown in Fig. 10). A predicted power constant $K_p = Po \cdot Re$ of 290 was close to the experimental value reported by Doucet et al. [65].

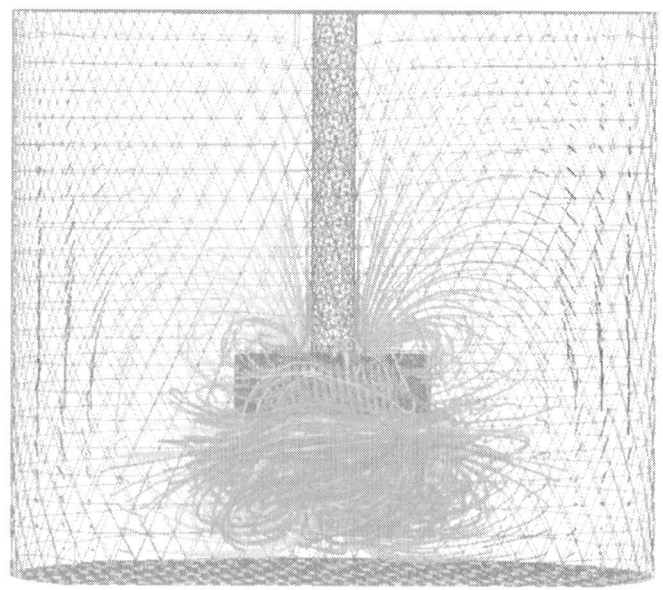

Figure 10: Predicted 3D flow lines in a vessel stirred by the VMI Rayneri batch high shear mixer at 300 rpm.

From Barailler et al. [94].

Mortensen et al. [95] recently reported 2D angular resolved PIV measurements in a custom-built batch HSM. To the authors' knowledge, this was the first PIV study of the turbulent flow characteristics in HSMs. Besides the general turbulent flow pattern (i.e. jets, recirculation and re-entrainment), strong stator slot vortices were experimentally captured. The highest turbulent intensity levels were found near and inside the stator slot vortex, where the mean velocity gradients were largest. Constant rotor tip speed (ND) scale-up criterion was suggested for batch high shear mixing processes that depend on macro-scale quantities, based on their experimental results that self similar flow pattern, mean velocity, strain rate and turbulence intensity fields would be maintained in geometrically similar HSM units.

MASS AND HEAT TRANSFER PROPERTIES

Mass and heat transfer are crucial for fast or mixing-sensitive chemical processes. Intensified mass and heat transfer can be achieved in HSMs, due to the fast surface renewal, large mass and heat transfer area and thin transfer sublayer in the highly turbulent flow. The actual mass and heat transfer rates depend on the equipment geometries and operating conditions that will influence the flow patterns, as well as the processed materials properties [96].

Liquid–Gas Mass Transfer

Deoxidization of water with nitrogen was utilized to investigate global mass transfer properties in in-line HSMs [97], [98] and [99]. The volumetric liquid–gas mass transfer coefficient (kxa) in HSMs was found in the same order of magnitude with that of rotating packed beds (RPBs) ($\sim 10^4$ mol/m^3 s). Results showed that kxa increased with an increase of the rotational speed, the gas flowrate and the liquid flowrate. Use of multiple rows of rotor–stator teeth promoted the liquid–gas mass transfer. Dimensionless analysis gave a correlation of $k_x a$ in the form of Eq. (11). When physical properties of the test fluids did not change obviously, this could be simplified as Eq. (12).

$$k_x a = \frac{\rho_L D_{ab}}{D} A_1 \cdot Re_G^{n_1} \cdot Re_L^{n_2} \cdot We^{n_3} \cdot Fr^{n_4} \tag{11}$$

$$k_x a = A_2 \cdot G^{m_1} \cdot L^{m_2} \cdot N^{m_3} \cdot D^{m_4} \tag{12}$$

where $k_x a$ is the volumetric liquid–gas mass transfer coefficient, ρ_L the liquid phase density and D_{ab} the binary diffusion coefficient; D is the nominal rotor diameter calculated as the mean diameter of the rotor–stator region; G, L, N are the gas flowrate, the liquid flowrate and the rotational speed of the HSM; Re_G and Re_L are gas and liquid phase Reynolds numbers, We the liquid phase Webber number and Fr the liquid phase Froude number. The correlations were verified within the rotor speed ranging from 300 to 1,450 rpm, the gas flowrate from 1.5 to 4.0 m³/h and the liquid flowrate from 0.75 to 2.0 m³/h.

Vapor–Liquid Heat Transfer

Heat transfer characteristics in in-line HSMs were evaluated using a steam–water system [100]. It is reported that the overall heat transfer coefficient increased with an increase of the rotor speed and the liquid flowrate. Dimensionless analysis gave a correlation of Nu in the form of Eq. (13). When physical properties of the operating fluids did not change obviously, this could be simplified as Eq. (14):

$$Nu = A_3 Re^{n_5} Pr^{n_6} \left(\frac{DN^2}{g} \right)^{n_7} \tag{13}$$

$$\alpha_0 = A_4 \Gamma m_5 V tip^{m_6} \tag{14}$$

where Nu, Re, Pr are the Nusselt number, Reynolds number and Prandtl number calculated based on liquid phase properties; D is the nominal rotor diameter calculated as the mean diameter of the rotor–stator region, N the rotational speed of the HSM, g the gravity acceleration, $\alpha 0$ the liquid phase heat transfer coefficient, Γ the mean liquid spray density, and V_{tip} the tip speed of the HSM. The correlations were verified within the rotor speed ranging from 400 to 1,400 rpm, and the water flowrate ranging from 0.1 to 0.5 m^3/h.

In the view of chemical reactors design, mass and heat transfer properties reflect the degree of process intensification and are of great significance for equipment evaluation. Especially, local transfer properties are useful for geometric optimizations of HSMs, in order to obtain a maximum selectivity of the target product in competing reaction systems with given reaction kinetics. However, it is still a technical challenge to measure detailed concentration and temperature distribution in complex HSMs for local transfer properties calculations. Transfer properties in other multiphase systems (e.g. liquid–liquid, liquid–solid or gas–liquid–solid) are also fundamental to chemical reactor engineering.

CONCLUSIONS AND OUTLOOKS

High shear mixers have been increasingly used for dispersed phase size reduction and mass and heat transfer intensification. To pave the way for the potentially promising applications in the field of chemical reaction processes, we here summarized major progress and key challenges on HSMs, as listed below.

1. **Liquid–Liquid Emulsification**: Due to the highly localized intense turbulence and shear in HSMs, the DSDs result from multiple breakage mechanisms and contain droplets with the size close to or below the Kolmogoroff microscale. CFD tools are necessary for predictions of the turbulence and shear in HSMs of different scales, as the geometric similarity is seldom obeyed in HSM scale-ups. The processing time plays an important role in rapidly coalescing emulsions and should be considered during scale-up. For practical HSM applications involving high fluid velocities, small daughter droplet sizes and high dispersed phase fractions, droplet size measurement becomes rather challenging. In these cases, CFD coupled population balance modeling with validated kernels are of

great help to disclose the droplet sizes evolution throughout the mixer.

2. **Solid–Liquid Suspension and Pharmaceutical Usage**: There is not much work on dispersion of powders compared to emulsions. Different mechanisms for particle breakage have been developed, but work is still needed to establish applicable correlations. HSMs have attracted much attention on the pharmaceutical polymorph transformation; however, it is fundamental to understand the mechanism that results in the polymorph transformation in HSMs so as to promote the potential substantial applications.

3. **Chemical Reactions and Transfer Properties Study**: HSMs have the potential to intensify mixing controlled chemical reactions due to their excellent micromixing efficiency. Feeding the reactants near the shear gap can be beneficial. Considering the circulation and possible bypassing flow characteristics in HSMs, the residence time distribution analyses are necessary for the diagnosis of the operational ills when operating them as chemical reactors. So far transfer properties have been studied at a global level, in which the reactors are treated as black boxes. Local transfer properties, which are useful for geometric optimizations of HSMs, have never been reported in the open literature. However, it is still a technical challenge to measure the detailed concentration and temperature distributions in complex HSMs for local transfer properties calculations. Transfer properties in other multiphase systems (e.g. liquid–liquid, liquid–solid or gas–liquid–solid), which are also important in chemical reactor engineering, need to be studied.

4. **Power Draw Characteristics**: Power draw characteristics in batch HSMs are similar to those in conventional stirred vessels, while power draw in in-line HSMs cannot be adequately described by a single Power number. The stator geometry has a significant influence on the power draw, as for both the batch and in-line HSMs. It is also valuable to study the power draw of HSMs with different rotor geometries, as well

as different rows and/or stages of rotor–stator units. Both the power draw and dispersion capability should be considered in the HSM selection, in order to develop an energy efficient process. Work still needs to be done to investigate the power draw characteristics of the HSMs when processing non-Newtonian fluids or multiphase fluids.

5. **Flow Pattern Analysis**: Laser-based EFD measurements require specially manufactured in-line HSMs with transparent front and/or circumference of the volute. Geometric complex makes it difficult to measure 3D flow field in the vicinity of the rotor of batch and in-line HSMs. CFD tools can be used for the single phase and multiphase flow simulations in HSMs, as well as the population balance coupled modeling. However, the prediction accuracy still needs to be improved, owing to the inherent weakness of the $k-$ turbulence model. Body-fitted computational grid with good quality and high resolution is essential to obtain fast convergence and acceptable accuracy. For the highly wall-bounded and strongly rotating unsteady flows in the geometrically complex HSMs, an assessment of different turbulence models should be conducted, e.g., using RNG $k-$ model to enhance accuracy for swirling flows, using the Reynolds Stress Model (RSM) to eliminate the unrealistic assumption of isotropic turbulence and using SST $k-$ model to ensure good performance in both the near-wall and far-field zones. However, RANS based turbulence models were reported to have an under-prediction tendency of turbulent kinetic energy. The large eddy simulation (LES), which has been attributed to significant successes in the complex flow simulations such as the stirred vessels [85], [86], [101],[102], [103], [104] and [105], turbomachineries [106], [107], [108] and [109] and cyclones [110], [111],[112] and [113], is a promising route to disclose more accurately the flow pattern and turbulent quantities in HSMs.

ACKNOWLEDGEMENT

The authors greatly acknowledged the support of NFSC (21076144) and the Special Funds for Major State Basic Research Program of China (2012CB720300).

REFERENCES

1. V.A. Atiemo-Obeng, R.V. Calabrese, Rotor–stator mixing devices, in: E.L. Paul, V.A.Atiemo-Obeng, S.M.Kresta (Eds.), Handbook ofIndustrialMixing: Science and Practice, John Wiley & Sons, Inc, New Jersey, 2004, pp. 479–505.

2. R.V. Calabrese, M.K. Francis, V.P. Mishra, S. Phongikaroon, Measurement and analysis of drop size in a batch rotor–stator mixer, in: Proc. 10th European Conference on Mixing, Delft, The Netherlands, 2000.

3. G.A. Padron, Effect of surfactants on drop size distributions in a batch, rotor–stator mixer, Ph.D. Dissertation, University of Maryland, College Park, Maryland, 2005.

4. S. Hall, M. Cooke, A. El-Hamouz, A.J. Kowalski, Droplet break-up by inline Silverson rotor–stator mixer, Chemical Engineering Science 66 (2011) 2068–2079.

5. Y.-F. Maa, C. Hsu, Liquid–liquid emulsification by rotor/stator homogenization, Journal of Controlled Release 38 (1996) 219–228.

6. J. Adler-Nissen, S.L. Mason, C. Jacobsen, Apparatus for emulsion production in small scale and under controlled shear conditions, Food and Bioproducts Processing 82 (2004) 311–319.

7. J.-P. Gingras, P.A. Tanguy, S. Mariotti, P. Chaverot, Effect of process parameters on bitumen emulsions, Chemical Engineering and Processing 44 (2005) 979–986.

8. K. Ouzineb, C. Lord, N. Lesauze, C. Graillat, P.A. Tanguy, T. McKenna, Homogenisation devices for the production of

miniemulsions, Chemical Engineering Science 61 (2006) 2994–3000.

9. U. El-Jaby, M. Cunningham, T. Enright, T.F.L. McKenna, Polymerisable miniemulsions using rotor–stator homogenisers, Macromolecular Reaction Engineering 2 (2008) 350–360.

10. J. Baldyga, W. Orciuch, L. Makowski, K. Malik, G. Özcan-Tas̨ kin, W. Eagles, G. Padron, Dispersion of nanoparticle clusters in a rotor–stator mixer, Industrial & Engineering Chemistry Research 47 (2008) 3652–3663.

11. J. Baldyga, W. Orciuch, L. Makowski, M. Malskibrodzicki, K. Malik, Break up of nano-particle clusters in high-shear devices, Chemical Engineering and Processing 46 (2007) 851–861.

12. J. Baldyga, L. Makowski, W. Orciuch, C. Sauter, H. Schuchmann, Deagglomeration processes in high-shear devices, Chemical Engineering Research and Design 86 (2008) 1369–1381.

13. N.G. Özcan-Tas̨ kin, G. Padron, A. Voelkel, Effect of particle type on the mechanisms of break up of nanoscale particle clusters, Chemical Engineering Research and Design 87 (2009) 468–473.

14. G. Padron, W.P. Eagles, G.N. Özcan-Tas̨ kin, G. McLeod, L. Xie, Effect of particle properties on the break up of nanoparticle clusters using an inline rotor–stator, Journal of Dispersion Science and Technology 29 (2008) 580–586.

15. A. Hassan, A. Hassan, High shear process for producing micronized waxes, US 20100125157A1, H.R.D. Corporation, 2010.

16. F.P.-H. Lee, R.W.-N. Wong, S.V. Kao, Process for preparing phase change inks, US 20070030322A1, Xerox Corporation, 2007.

17. S. Shelley, High-shear mixers: still widely misunderstood, Chemical Engineering Progress 103 (2007) 7–12.

18. A. Hassan, E. Bagherzadeh, R.G. Anthony, G. Borsinger,

A. Hassan, High shear system for the production of chlorobenzene, US 20100183486A1, H.R.D. Corporation, 2010.

19. A. Hassan, E. Bagherzadeh, R.G. Anthony, G. Borsinger, A. Hassan, System for making linear alkylbenzenes, US 20100266465A1, H.R.D. Corporation, 2010.

20. A. Hassan, E. Bagherzadeh, R.G. Anthony, G. Borsinger, A. Hassan, System and process for production of toluene diisocyanate, PCT/US 2008067835, H.R.D. Corporation, 2008.

21. A. Hassan, E. Bagherzadeh, R.G. Anthony, G. Borsinger, A. Hassan, System and process for production of toluene diisocyanate, US 20110027147A1, H.R.D. Corporation, 2011.

22. A. Hassan, E. Bagherzadeh, R.G. Anthony, G. Borsinger, A. Hassan, System and process for the production of aniline and toluenediamine, US 007750188B2, H.R.D. Corporation, 2010.

23. E. Vanzo, L.S. Smith, Method of sizing monomer droplets for suspension polymerization to form small particles, US 4071670, Xerxo Corporation, 1978.

24. J.T. Davies, A physical interpretation of drop sizes in homogenizers and agitated tanks, including the dispersion of viscous oils, Chemical Engineering Science 42 (1987) 1671–1676.

25. M.K. Francis, The development of a novel probe for the in situ measurement of particle size distributions, and application to the measurement of drop size in rotor–stator mixers, Ph.D. Dissertation, University of Maryland, College Park, Maryland, 1999.

26. S. Phongikaroon, Drop size distribution for liquid-liquid dispersions produced by rotor–stator mixers, Ph.D. Dissertation, University of Maryland, College Park, MD, 2001.

27. R.V. Calabrese, M.K. Francis, K.R. Kevala, V.P. Mishra, G.A. Padron, S. Phongikaroon, Fluid dynamics and emulsification

in high shear mixers, in: Proc. 3rd World Congress on Emulsions, Lyon, France, 2002.

28. S. Phongikaroon, R.V. Calabrese, K. Carpenter, Elucidation of polyurethane dispersions in a batch rotor–stator mixer, Journal of Coatings Technology and Research 1 (2004) 329–335.

29. J.T. Davies, Drop sizes of emulsions related to turbulent energy dissipation rates, Chemical Engineering Science 40 (1985) 839–842.

30. A.R. Khopkar, L. Fradette, P.A. Tanguy, Emulsification capability of a dual shaft mixer, Chemical Engineering Research and Design 87 (2009) 1631– 1639.

31. Y.N. Averbukh, A.O. Nikoforov, N.M. Kostin, A.V. Korshakov, Computation of dispersion for emulsions formed in a rotor–stator unit, Journal of Applied Chemistry of the USSR 61 (1988) 396–397. 40 J. Zhang et al. / Chemical Engineering and Processing 57–58 (2012) 25–41

32. N. Thapar, Liquid–liquid dispersions from in-line rotor–stator mixers, Ph.D. Dissertation, Cranfield University, Cranfield Bedfordshire, 2004.

33. K.R. Kevala, K.T. Kiger, R.V. Calabrese, Single pass drop size distributions in an inline rotor–stator mixer, in: APS Division of Fluid Dynamics 58th Annual Meeting (DFD05), Chicago, IL, 2005.

34. S. Hall, M. Cooke, A.W. Pacek, A.J. Kowalski, D. Rothman, Scaling up of Silverson rotor–stator mixers, Canadian Journal of Chemical Engineering 89 (2011) 1040–1050.

35. A.W. Pacek, C.C. Man, A.W. Nienow, On the Sauter mean diameter and size distributions in turbulent liquid/liquid dispersions in a stirred vessel, Chemical Engineering Science 53 (1998) 2005–2011.

36. C. Angle, H. Hamza, Predicting the sizes of toluene-diluted heavy oil emulsions in turbulent flow Part 2: Hinze–Kolmogorov based model adapted for increased oil fractions and energy dissipation in a stirred tank, Chemical Engineering Science 61 (2006) 7325–7335.

37. G.W. Zhou, S.M. Kresta, Correlation of mean drop size and minimum drop size with the turbulence energy dissipation and the flow in an agitated tank, Chemical Engineering Science 53 (1998) 2063–2079.

38. R.V. Calabrese, T.P.K. Chang, P.T. Dang, Drop breakup in turbulent stirred-tank contactors Part I: effect of dispersed-phase viscostiy, AIChE Journal 32 (1986) 657–666.

39. R. Shinnar, On the behaviour of liquid dispersions in mixing vessels, Journal of Fluid Mechanics 10 (1961) 259–275.

40. D.E. Leng, R.V. Calabrese, Immiscible liquid–liquid systems, in: E.L. Paul, V.A. Atiemo-Obeng, S.M. Kresta (Eds.), Handbook ofIndustrial Mixing: Science and Practice, John Wiley & Sons, Inc, New Jersey, 2004, pp. 639–746.

41. M. Nishikawa, F. Mori, S. Fujieda, Average drop size in a liquid-liquid phase mixing vessel, Journal of Chemical Engineering of Japan 20 (1987) 82–88.

42. F. Theron, N.L. Sauze, A. Ricard, Turbulent liquid–liquid dispersion in Sulzer SMX mixer, Industrial & Engineering Chemistry Research 49 (2010) 623–632.

43. T. Lemenand, Droplets formation in turbulent mixing oftwo immiscible fluids in a new type of static mixer, International Journal of Multiphase Flow 29 (2003) 813–840.

44. G. Farzi, M. Mortezaei, A. Badiei, Relationship between droplet size and fluid flow characteristics in miniemulsion polymerization of methyl methacrylate, Journal of Applied Polymer Science 120 (2011) 1591–1596.

45. J.-P. Gingras, L. Fradette, P. Tanguy, J. Bousquet, Inline bitumen emulsification using static mixers, Industrial & Engineering Chemistry Research 46 (2007) 2618–2627.

46. R. Thakur, C. Vial, K. Nigam, E. Nauman, G. Djelveh, Static mixers in the process industries – a review, Chemical Engineering Research and Design 81 (2003) 787–826.

47. P.D. Berkman, R.V. Calabrese, Dispersion of viscous liquids by turbulent flow in a static mixer, AIChE Journal 34 (1988) 602–609.

48. A.N. Kolmogoroff, The breakup of droplets in a turbulent stream, Doklady Akademii Nauk 66 (1949) 825–828.

49. J.O. Hinze, Fundamentals of the hydrodynamics mechanism of splitting in dispersion processes, AIChE Journal 1 (1955) 289–295.

50. H. Karbstein, H. Schubert, Developments in the continuous mechanical production of oil-in-water macro-emulsions, Chemical Engineering and Processing 34 (1995) 205–211.

51. G.K. Patterson, E.L. Paul, S.M. Kresta, A.W. Etchells, Mixing and chemical reactions, in: E.L. Paul, V.A. Atiemo-Obeng, S.M. Kresta (Eds.), Handbook of Industrial Mixing: Science and Practice, John Wiley & Sons, Inc, New Jersey, 2004, pp. 755–867.

52. C. Guang-wen, S. Yun-hua, C. Jian-ming, C. Han, Preparation of nano-CaCO3 by rotor–stator reactor, Chemical Industry and Engineering Progress 24 (2005) 545–549.

53. S. Yun-hua, L. Jing-jing, C. Jian-ming, Y. Jian-jun, Preparation of basic magnesium sulfate whiskers by rotor–stator reactor and hydrothermal method, CIESC Journal 60 (2009) 756–761.

54. J.R. Bourne, J. Garcia-Rosas, Rotor–stator mixers for rapid micromixing, Chemical Engineering Research and Design 64 (1986) 11–17.

55. J.R. Bourne, M. Studer, Fast reactions in rotor–stator mixers of different size, Chemical Engineering and Processing 31 (1992) 285–296.

56. G. Chu, Y. Song, H. Yang, J. Chen, H. Chen, Micromixing efficiency of a novel rotor–stator reactor, Chemical Engineering Journal 128 (2007) 191–196.

57. Z. Zhan-yuan, M. Jian, G. Zheng-ming, Micromixing characteristics of a continuous rotor–stator mixer, Journal of Beijing University of Chemical Technology 35 (2008) 4–7.

58. Y. Lei, L. Zhi-peng, G. Zheng-ming, Micromixing characteristics in a continuous rotor–stator mixer, Journal of Beijing University of Chemical Technology 37 (2010) 1–4.

59. L. Jun-bo, Mixcromixing characteristics and application in rotor–stator reactor, M.E. Thesis, Beijing University of Chemical Technology, Beijing, 2008.

60. D.A.R. Brown, P.N. Jones, J.C. Middleton, G. Papadopoulos, E.B. Arik, Experimental methods, in: E.L. Paul, V.A. Atiemo-Obeng, S.M. Kresta (Eds.), Handbook of Industrial Mixing: Science and Practice, John Wiley & Sons, Inc, New Jersey, 2004, pp. 145–250.

61. G. Ascanio, B. Castro, E. Galindo, Measurement of power consumption in stirred vessels – A review, Chemical Engineering Research and Design 82 (2004) 1282–1290.

62. M.F. Edwards, M.R. Baker, J.C. Godfrey, Mixing of liquids in stirred tanks, in: N. Harnby, M.F. Edwards, A.W. Nienow (Eds.), Mixing in the Process Industries, second edition, Butterworth-Heinemann Ltd, Oxford, 1992, pp. 137–159.

63. G.A. Padron, Measurement and comparison of power draw in batchrotor–stator mixers, M.S. thesis, University of Maryland, College Park, Maryland,2001.

64. K.J. Myers, M.F. Reeder, D. Ryan, Power draw of a high-shear homogenizer,Canadian Journal of Chemical Engineering 79 (2001) 94–99.

65. L. Doucet, G. Ascanio, P. Tanguy, Hydrodynamics characterization ofrotor–stator mixer with viscous fluids, Chemical Engineering Research andDesign 83 (2005) 1186–1195.

66. A. Khopkar, L. Fradette, P. Tanguy, Hydrodynamics of a dual shaft mixer withNewtonian and non-Newtonian fluids, Chemical Engineering Research andDesign 85 (2007) 863–871.

67. S.Kohler, H. Bord, P.Walzel, The blending efficiency of an intermeshing helicalribbon and screw impeller combination, in: Proceedings of 12th EuropeanConference on Mixing, Bologna, Italy, 2006.

68. A.B. Merzner, R.E. Otto, Agitation of non-Newtonian fluids,

AIChE Journal 3 (1957) 3–10.

69. F. Rieger, V. Novák, Power consumption of agitators in highly viscous nonNewtonian liquids, Chemical Engineering Research and Design 51 (1973) 105–111.

70. S. Nagata, Mixing: Principles and Applications, Halsted Press, New York, 1975.

71. K. Guerinik, Agitation and aeration of viscoelastic fermentation broths, Ph.D. Dissertation, University of Laval, Quebec, Canada, 1990.

72. E. Brito de La Fuente, Mixing of rheological complex fluids with helical ribbon and helical screw ribbon impellers, Ph.D. Dissertation, University of Laval, Quebec, Canada, 1992.

73. J.H. Rushton, E.W. Costich, H.J. Everett, Power characteristics of mixing impellers, Chemical Engineering Progress 46 (1950) 467–476.

74. M.Cooke,J. Naughton,A.J.Kowalski,Asimplemeasuremen tmethodfordetermining the constants for the prediction of turbulent power in a Silverson MS 150/250 in-line rotor–stator mixer, in: 6th International Symposium on Mixing in Industrial Process Industries-ISMIP VI, Ontario, Canada, 2008.

75. A. Kowalski, An expression for the power consumption of in-line rotor–stator devices, Chemical Engineering and Processing 48 (2009) 581–585.

76. A.J. Kowalski, M. Cooke, S. Hall, Expression for turbulent power draw of an in-line Silverson high shear mixer, Chemical Engineering Science 66 (2011) 241–249.

77. M. Cooke, T.L. Rodgers, A.J. Kowalski, Power consumption characteristics of an in-line Silverson high shear mixer, AIChE Journal (2011), http://dx.doi.org/10.1002/aic.12703.

78. N.G. Özcan-Taş kin, D. Kubicki, G. Padron, Power and flow characteristics of three rotor–stator heads, Canadian Journal of Chemical Engineering 89 (2011) 1005–1017.

79.   A.T. Utomo, M. Baker, A.W. Pacek, Flow pattern, periodicity and energy dissipation in a batch rotor–stator mixer, Chemical Engineering Research and Design 86 (2008) 1397–1409.

80.   Fluent, User's Manual to Fluent 6.3, Fluent Inc., Central Resource Park, 10 Cavendish Court, Lebanon, USA, 2006.

81.   E.M. Marshall, A. Bakker, Computational fluid mixing, in: E.L. Paul, V.A. Atiemo-Obeng, S.M. Kresta (Eds.), Handbook of Industrial Mixing: Science and Practice, John Wiley & Sons, Inc, New Jersey, 2004, pp. 257–341.

82.   K. Ng, N. Fentiman, K. Lee, M. Yianneskis, Assessment of sliding mesh CFD predictions and LDA measurements of the flow in a tank stirred by a Rushton impeller, Chemical Engineering Research and Design 76 (1998) 737– 747.

83.   K. Ng, M. Yianneskis, Observations on the distribution of energy dissipation in stirred vessels, Chemical Engineering Research and Design 78 (2000) 334–341.

84.   Z. Jaworski, B. Zakrzewska, Modelling of the turbulent wall jet generated by a pitched blade turbine impeller: the effect of turbulence model, Chemical Engineering Research and Design 80 (2002) 846–854.

85.   S.L. Yeoh, G. Papadakis, M. Yianneskis, Numerical simulation ofturbulent flow characteristics in a stirred vessel using the LES and RANS approaches with the sliding/deforming mesh methodology, Chemical Engineering Research and Design 82 (2004) 834–848.

86.   B. Murthy, J. Joshi, Assessment of standard k–, RSM and LES turbulence models in a baffled stirred vessel agitated by various impeller designs, Chemical Engineering Science 63 (2008) 5468–5495.

87.   K.R. Kevala, Sliding mesh simulation of a wide and narrow gap inline rotor–stator mixer, M.S. Thesis, University of Maryland, College Park, MD, 2001.

88.   A.W. Pacek, M. Baker, A.T. Utomo, Characterisation of flow pattern in a rotor stator high shear mixer, in: Proc. 6th European

Congress on Chemical Engineering (ECCE-6), Copenhagen, 2007.

89. A.T. Utomo, M. Baker, A.W. Pacek, The effect of stator geometry on the flow pattern and energy dissipation rate in a rotor–stator mixer, Chemical Engineering Research and Design 87 (2009) 533–542.

90. A.T. Utomo, Flow pattern and energy dissipation rates in batch rotor–stator mixers, Ph.D. Dissertation, The University of Birmingham, Edgbaston, Birmingham B15 2TT, 2009.

91. A. Utomo, M. Baker, A.W. Pacek, The effect of stator geometry in the flow pattern and energy dissipation rate in a rotor–stator mixer, in: 13th European Conference on Mixing, London, UK, 2009.

92. A. Bakker, H.E.A. Van den Akker, The use of profiled axial flow impellers in gas–liquid reactors, in: Fluid Mixing IV, IChemE, Bradford, U.K, 1990.

93. A.W. Nienow, On impeller circulation and mixing effectiveness in the turbulent flow regime, Chemical Engineering Science 52 (1997) 2557–2565.

Chapter 4

Roles of Drag Reducing Polymers in Single- and Multi-Phase Flows

A. Abubakar, T. Al-Wahaibi, Y. Al-Wahaibi, A.R. Al-Hashmi, and A. Al-Ajmi

Department of Petroleum and Chemical Engineering, Sultan Qaboos University, Al-Khoud 123, Oman

ABSTRACT

It has become a well-known fact that finding sustainable solutions to the unavoidable high pressure losses accompanying pipeline flows to increase the pumping capacity without necessarily adding more pump stations is inevitable. Polymers, as one of the drag reducing agents which have been found to offer such an economic relieve, is the most widely investigated and most often employed in industries because they can produce drag reduction up to 80% when they are added in minute concentrations. In addition, polymer additives modify the flow configurations of multiphase flows to such an extent that stratification of individual phases is enhanced thereby making the separation of the phases at the fluid destination much easier. The achievements so far made and the challenges facing the use of polymers as drag reducers in turbulent single and multiphase flows are comprehensively reviewed. This review discusses the experimental studies of the effects of polymer additives in turbulent flows, the analytical studies, and the proposed models as well as the suggested mechanisms that explain the drag reduction. Likewise, specific areas of interest in the review include phenomena of drag reduction by polymers, factors influencing the effectiveness of the drag reducing polymers, methods of injecting the polymers into the base fluids, degradation of the polymers and industrial applications of polymers as drag reducing agents. The current and future research interests are also addressed. Although finding reveals that there are quite a lot of research in this area, most of the experimental and theoretical works are devoted to single phase flows while the remaining ones are mostly directed towards gas–liquid flows except in very recent time when investigation into the use of polymers in liquid–liquid flows is being focused. Despite this voluminous works on drag reducing polymers, there are no universally accepted models and hence the mechanisms of drag reductions by polymers.

INTRODUCTION

The concept of drag (or pressure drop) reduction in pipe flow using drag reducing agents (DRAs) has generated practical engineering interest because of these agents' abilities to reduce pumping power and increase piping system capacity. Recently, some drag reduction studies in multiphase flows revealed that apart from pressure drop, DRAs can also affect the spatial distribution of the fluids in the pipe and the boundaries between different flow patterns (Oliver and Young, 1968, Greskovich and Shrier, 1971, Virk, 1975, Al-Sarkhi and Hanratty, 2001a and Al-Sarkhi and Hanratty, 2001b). Such areas of interest where these additives are particularly gaining tremendous attention are oil production and transportation pipelines, and district heating and cooling. The additives causing drag reduction can be divided into five categories: polymers, surfactants, fibres, micro-bubbles and compliant coating. This review presents only the use of polymers as drag reducers in turbulent single and multiphase flows as they have been most employed at industrial level due to their advantage of being applied in very small quantities when compared to surfactants and fibres.

Drag reducing polymers (DRPs) are long chain, ultra-high molecular weight (typically ranging from 1 to 10 millions) polymers which can be water- and/or oil-soluble. With the higher molecular weight polymers giving better drag reduction performance, only parts-per-million levels of the polymers in the working fluid suppress the formation of turbulent bursts in the buffer region, and in turn suppress the formation and propagation of turbulent eddies (Fig. 1). This causes the hydraulic energy provided by the pumps to be more directed to moving the fluid down the pipeline rather than being used for a chaotic and random motion.

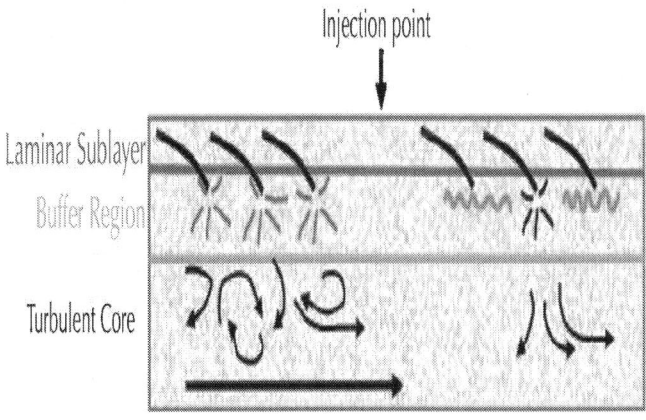

Figure 1: Illustration of pipeline turbulent flow regions.

The formal discovery of drag reduction using polymers is attributed to Toms (1948) during experiments on the degradation of polymers in pump. Since then, a lot of research for both practical and fundamental purposes has been going on to further understand and adequately utilize this phenomenon resulting probably in thousands of papers in both single and multiphase phase flows. Due to the huge number of publications in the subject, several review articles are currently available for both single phase flow (see for example, Lumley, 1969 and recently Gyr and Bewersdorff, 1995 and Manfield et al., 1999) and multiphase flow (Jubran et al., 2005 and Al-Sarkhi, 2010).

This review aims to provide insight into the subject with an emphasis on the recent progress in both single and multiphase flows. In fact, it is intended to bring a sharp focus into previous research efforts in this emerging field of DRPs' use to reduce drag and modify flow patterns, methodology, the knowledge gaps, and the future trends of research.

DRAG REDUCTIONS IN SINGLE PHASE FLOW EXPERIMENTS

The most extensive research works on the use of DRPs that have been documented involve single liquid phase flow. Although the application of these polymer additives on gas–liquid turbulent flow is relatively much published, other multiphase flow, especially liquid–liquid flow, has received the least attention (Al-Wahaibi et al., 2007 and Al-Sarkhi, 2010).

As early as 1960, the effect of drag reducing agents on fluid flow has been under investigation (for extensive review see Lumley, 1969). However, one of the earliest well established and generally accepted study was the one carried out by Virk (1975). He investigated the effect of DRP on the mean velocity profile by introducing the idea of an elastic sub-layer which exists between the viscous sub-layer and the outer Newtonian region of the flow.

The main finding was the increase in wall layer thickness in the presence of DRP which extended the wall layer velocity profile, a phenomenon which was generally accepted in subsequent investigations (Pinho and Whitelaw, 1990).

Drag Reduction and Maximum Drag Reduction

All previous studies including Virk (1975) work concluded that drag reduction effect occurs only in turbulent regime where the Reynolds number is above 2300 in pipe flows. This is because no significant difference was observed in the skin friction between laminar flow of Newtonian fluids and laminar flow of dilute polymer solutions.

The effectiveness of the drag reducing polymers is determined by the percentage of drag reduction (DR) in a flowing fluid which is usually expressed quantitatively as follows, though equivalent expressions in other dimensional quantities can be formulated.

$$DR = \frac{\Delta P_{without\,DRP} - \Delta P_{with\,DRP}}{\Delta P_{without\,DRP}}$$

(1)

where $P_{without\,DRP}$, pressure drop in the absence of DRP; $P_{with\,DRP}$, pressure drop in the presence of DRP.

Drag reduction can also be represented in terms of Reynolds number and friction factor. In a more insightful representation, the plot of these dimensionless quantities of polymer solution in terms of Prandtl–Karman (P–K) parameters such as $1/\sqrt{f}$ against $Re/\sqrt{f}$ shows that above a certain Reynolds number, the friction factor falls below that for the case of only pure solvent flow. Generally, there is a separate relationship between these parameters for laminar and turbulent regimes in Newtonian fluids, like the polymer solvent, shown in Eqs. (2) and (3).

$$\frac{1}{\sqrt{f}} = \frac{Re\sqrt{f}}{16} \quad \text{Laminar flow}$$

$$(2)$$

$$\frac{1}{\sqrt{f}} = 4\log_{10} Re\sqrt{f} - 0.4 \quad \text{Turbulent flow}$$

$$(3)$$

Two main types of drag reduction can be distinguished based on the region where they display drag reduction (Wang et al., 2011). The first is drag reduction by very dilute solutions that display onset drag reduction in the fully developed turbulence region where drag reduction only occurs above an onset Reynolds number and friction factor decreases below that for ordinary Newtonian turbulent flow. The second is drag reduction by more concentrated polymer solutions that display drag reduction in the extended laminar region at low Reynolds numbers. Both types of drag reduction increase with flow rate until a critical wall shear stress is attained. At such shear stress, the rate of polymer degradation is higher than the rate at which polymer is replenish (Hoyt, 1989). Drag reduction can equally be regarded as fan type (or Type A) and ladder type (or Type B) depending on the behaviours of the polymer solutions towards concentration of salt added to the solution (Virk, 1975). Type A is characterized by the presence of an onset stress above which DR starts whereas type B is characterized by the extension of the laminar flow line into the turbulent region (Hershey and Zakin, 1967 and Virk and Wagger, 1989).

Interestingly, the initiation of drag reduction by polymer additives occurs at the same Reynolds number regardless of the polymer concentration. However, it occurs at lower Reynolds number when the molecular weights of the polymers are increased (Virk, 1975). In addition, only very small drag reducing polymer concentrations in ppm are needed to create a significant change in drag reduction (Hoyt, 1989 and Warholic et al., 1999). Warholic et al. (1999) showed that a significant drag reduction can be achieved with only 0.25 ppm polymer concentration. Experimental results have also shown that drag reduction rapidly increases as the concentration of the polymer is increased until it levels off with the concentration at maximum drag reduction (Sifferman and Greenkorn, 1981). This is the observation of most investigators but some later findings have revealed that further increase in concentration will slightly reduce the drag reduction as shown in Fig. 2 (Hoyt, 1989).

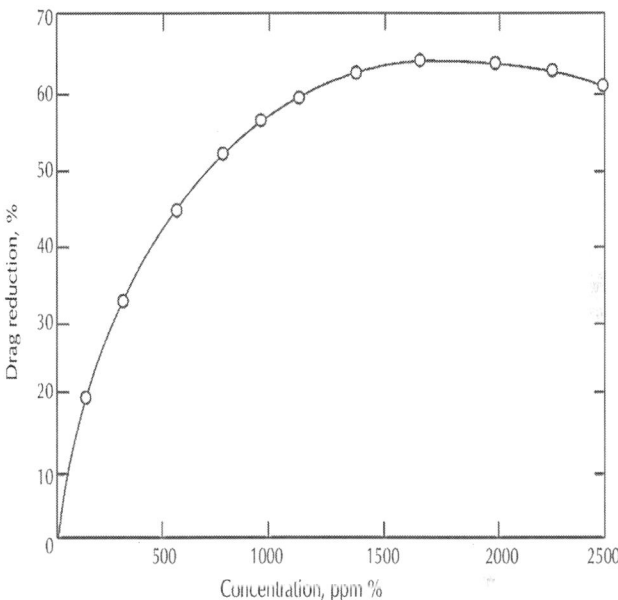

Figure 2: Drag reduction as a function of DRP concentration for a pipe flow of Reynolds number of 14,000 (Hoyt, 1989).

Meanwhile, if the Reynolds number is increased at fixed polymer concentration, studies like White and Mungal (2008) have revealed that the drag reduction increases quickly until it reaches the maximum, forming what is known as maximum drag reduction (MDR) asymptote. This trend is found to be applicable to a very wide range of polymer solutions (Fig. 3). The maximum drag reduction tends to increase with increase in pipe diameter. One the other hand, drag reduction (in the region below the maximum) decreases with increase in pipe diameter (Interthal and Wilski, 1985). The representation of maximum drag reduction in terms of $1/\sqrt{f}$ against $Re/\sqrt{f}$ plot is given by;

$$\frac{1}{\sqrt{f}} = 19 \log_{10} \mathrm{Re} \sqrt{f} - 3.24$$

(4)

The behaviour of polymer drag reduction can also be represented in terms of velocity profiles using polymer-induced modification to the law-of-the-wall according to the work of Virk (1975) as presented in Fig. 4. Von Karman (1930) published a law for flow in a tube called law-of-the-wall which states that the average velocity of a turbulent flow at a certain point is proportional to the logarithm of the distance from that point to the "wall", or the boundary of the fluid region. In the absence of drag reduction, the Newtonian velocity profile for the turbulent boundary inner layer consists of laminar (viscous) sub-layer which exists below the region where this law is applicable and log-law region for the turbulent core where the law-of-the-wall is valid. Essentially, this log-law region refers to the law-of-the-wall, and the velocity profiles of these two distinct regions for Newtonian fluids are respectively represented by Eqs.(5) and (6).

$$U^+ = y^+, \quad y^+ < 11.6$$

(5)

$$U^+ = 2.5 \ln y^+ + 5.5, \quad y^+ > 11.6$$

(6)

where U^+ is the velocity scale parallel to the pipe and y^+ is the length scale normal to the wall. The superscript "+" indicates dimensionless turbulence scales which are obtained as follows:

$$U^+ = \frac{U}{U_T}, \quad y^+ = \frac{y U_T}{\vartheta} \quad \text{and} \quad U_T = \sqrt{\frac{\tau_w}{\rho}}$$

where U is the velocity parallel to the pipe, y is the distance from the wall, UT is the turbulent wall (or friction) velocity, τ_w is wall shear stress while t and ρ are the kinematic viscosity and the density of the fluid respectively.

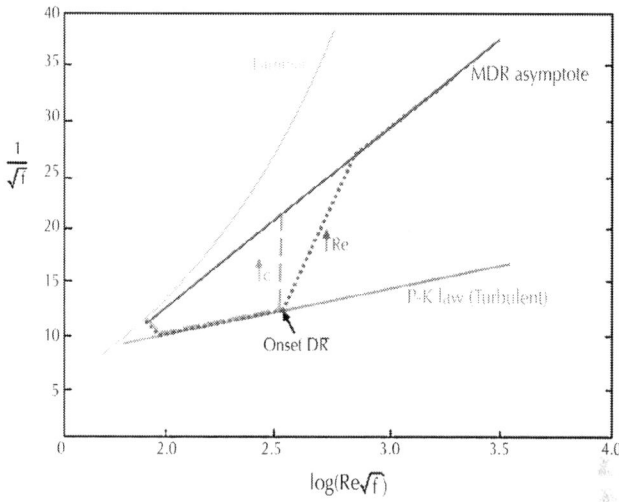

Figure 3: A schematic illustration of onset and different trajectories of polymer drag reduction. The dashed line represents the case in whichRe is fixed (at the value when the onset of drag reduction is first observed) and polymer concentration, C, is increased. The dotted line represents the case in which C is fixed and Re is increased (White and Mungal, 2008).

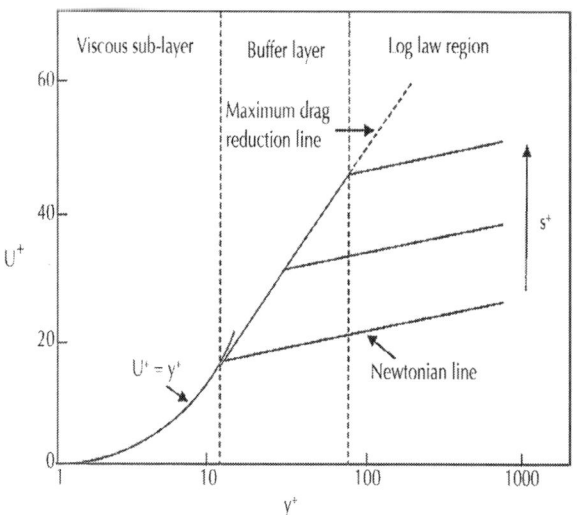

Figure 4: Velocity profile of Virk's phenomenon model (Virk, 1975).

In the presence of drag reducing polymer, the Newtonian law-of-the-wall is shifted upward by an amount S^+ with no change in the slope as shown in Fig. 4. Thus, Eq. (5) is modified to give Eq. (7) which accounts for non-Newtonian behaviour of the polymer solution (pseudoplastic or shear-shinning fluid) at the turbulent state (Larson, 2003).

$$U^+ = 2.5 \ln y^+ + 5.5 + S^+ \tag{7}$$

$$U^+ = 11.7 \ln y^+ - 17.0 \tag{8}$$

When this happens, intermediate buffer layer which is equivalent to Virk's elastic sub-layer is introduced as given in Eq. (8), generating a maximum drag reduction segment which connects the laminar sub-layer curve to the shifted law-of-the-wall line for the turbulent core. This will occur for the whole flow field and therefore Eq. (8) represents the ultimate velocity profile for maximum drag reduction by polymers.

Turbulence Measurements

Turbulence can be described as a fluctuating and chaotic fluid motion which manifests when nonlinear inertia effects dominate over viscous effects. Generally, turbulent flow exhibits random fluctuations of linear quantities (velocity, temperature, pressure and density) around their mean value. However, as far as drag reducing turbulent flows are concerned, the characterizing quantity that is of paramount importance is the velocity. Fluctuating velocity components which depend on their frequencies and amplitudes imparted transverse motions to the particles of the liquid that move in the main (axial) flow direction causing mass transfer between adjacent layers of the liquid which results in continuous stirring of the fluid. Also, the fact that energy is needed to overcome viscous drag as well as to maintain the continuous stirring of the fluid (Fig. 5a) shows that turbulent flow is characterized by a higher consumption of energy at a given velocity. However, the presence of a polymer additive theoretically leads to the thickening of the laminar sublayer (known as partial laminarization) of the flow, which causes an increase in the mean velocity (Fig. 5b).

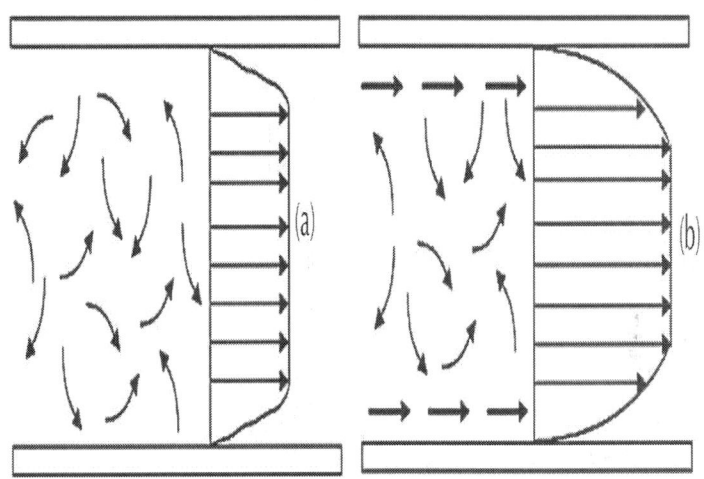

Figure 5: Velocity profiles of the turbulent flow of (a) a pure liquid and (b) a liquid that contains a polymer additive (Nesyn et al., 2012).

A complete insight into the dynamics of turbulence is still lacking because non-linear equations governing turbulent flow have not been clearly and satisfactorily analyzed. Turbulence is mostly studied from statistical point of view and to predict with accuracy the detailed evolution of a turbulent flow is in principle impossible. It must be stressed however that since turbulence affects technical issues such as drag reduction, the accurate prediction of its evolution and measurement of its coherent structures become important. Basically, turbulence measurement in fluid flows involving drag reduction is largely connected with the measurement of flow velocities. Early turbulence research has been complemented by the point measurement technique of Hot Wire Anemometry (HWA). This is in addition to the use of flow visualization measurements with the aid of dyes and particles as well as hydrogen bubbles generated from electrolysis in water. Up till today HWA system is employed majorly in form of Constant

Temperature Anemometry (CTA) implementation, which uses heat loss from a heated wire or film to measure velocity. Despite the advantages of CTA which include excellent spatial and temporal resolution, use of multi-probes, one- or multi-dimensional flexibility and relatively cheapness, the difficulties in using these intrusive methods include reversing flows, vortices, and highly turbulent flows. In addition, intrusive probes are subject to non-linearity that requires calibration; contamination of the probe like dirt and bubbles; sensitivity to multi-variable effects such as temperature and humidity; and breakage among other problems.

Fortunately, non-intrusive measurements of turbulent flow were realized in the mid-1960s utilizing lasers. Yeh and Cummins (1964) developed Laser-Doppler Anemometry (LDA) or Velocimetry (LDV). In contrast to the CTA, the LDA serves as a nearly ideal transducer in the sense that its output is linear, has no noise and no need for calibrations. Moreover, high frequency response can be obtained and the velocity is independently measured of other flow variables. The LDA technique has witnessed significant advancements in terms of optical methods such as fibres, and advanced signal processing techniques and software development in the last four decades (Clifford and French, 1993 and Jensen, 2004).

Interestingly, the LDA method has also been extended to the Phase Doppler technique for the measurement of particle and bubble size along with velocity. Additionally, the rapid development within lasers and camera technology has opened the possibility for qualifying (flow visualization) and later on quantifying whole flow field measurement. Acoustic Doppler velocitimeters are the modified versions of LDA that use the same principles. However, the need for clear flows (non-opaque), good laser light intensity, considerations of tracer particle (signal) drop-out (i.e. may not be a continuous signal) as well as safety issue and expensive nature are the challenges facing the use of LDA. Another turbulent flow measuring instrument that was developed after LDA is the Particle Image Velocimetry (PIV) which has become one of the most popular instruments for flow measurements in numerous applications. Specifically, it is now on record that LDA and PIV are currently the most commonly used and commercially available diagnostic techniques to measure fluid flow velocity (Clifford and French, 1993).

PIV uses the principle of change in position of tracer particles between two video or photo images to calculate velocity. It has also been suggested that with the modern developments of camera and laser technology, as well as PIV software, the performance of the PIV systems and their applicability to difficult flow measurements will continue to improve. In addition to the instantaneous measurement of the flow, a time resolved measurement is now possible with high frequency lasers and high frame rate cameras. Although both LDA and PIV have similar disadvantages, the modified PIV system that includes a second camera has the ability to measure particle and bubble sizing in turbulent flows as well (Clifford and French, 1993 and Jensen, 2004).

Many experimental research involving drag reduction in turbulent flows have applied these techniques.Toonder (1995) used LDA technique to measure turbulence in pipe flow of water and dilute polymer solutions. Hibberd et al. (1982), Harder and Tiederman (1991), Ptasinski et al. (2001), Warholic et al. (2001), Kim et al.

(2004) and White et al. (2004) reported the use of PIV while Zeybek (2005) and Kim and Sirviente (2005) used Laser Doppler Velocimetry (LDV) to measure turbulence in flow. Hou et al. (2006)studied polymer drag reduced turbulent boundary layer using Planar Laser Induced Fluorescence (PLIF) combined with Particle Image Velocimetry, though PLIF is used particularly for concentration and temperature measurements in heat transfer and mixing studies. It should be noted that what are usually employed for turbulence measurements in the field are rotary current metres, electromagnetic current metres and acoustic Doppler instruments.

Parameters Influencing the Performance of DRPs in Single Phase Flow

Apart from the effect of polymer concentrations on drag reductions in single-phase flow, there has been ongoing interest in investigating all the possible factors that may influence the performance of the drag reducing polymers and drag reduction. Detailed discussion of the effect of these various parameters is summarized in the following sections:

Effect of Channel Size and Geometry

One of many fascinating aspects of these non-Newtonian drag-reducing flows is the so-called 'diameter effect' which is the additional dependence of the friction factor on the pipe diameter besides the dependence on the Reynolds number which is the only parameter needed to define the friction factor for Newtonian fluids in smooth pipes. Although the influence of pipe diameter on the performance of DRP was studied by different investigators (Virk, 1975, Interthal and Wilski, 1985 and Karami and Mowla, 2012), how the amount of drag reduction (%DR) is affected by change in the internal diameter (ID) of the conduit or pipe had been and is still a source of controversy because the positions of researchers on this issue do not agree. However, Interthal and Wilski (1985) published one of the most comprehensive results on the influence of DRPs by factors such as pipe diameters. They reported that drag reduction increased from 66% at 3-mm ID to a peak of 80% at 14-mm ID and then declined to 76% at the highest 30-mm ID. This result confirmed that there is lack of consistency in the variation of drag reduction with pipe diameters. Another similar study is the investigation carried out by Karami and Mowla (2012). They investigated the drag reduction of three different polymer solutions with the same concentration of 200 ppm and at 29 °C in two rough galvanized iron pipes consisting of 0.0254 and 0.0127-m IDs. They observed that the DR decreases with increase in pipe diameter for all the polymer solutions.

As shown above, most of the DRP studies reported in the literature were concerned with pipe flow and to some extent with straight channel. However, other geometries such coiled tubes have also been studied (Shah and Zhou, 2003, Shah and Zhou, 2009, Zhou et al., 2006, Fox et al., 2010 and Kamel, 2011). In all these investigations, it is generally observed that drag reduction is more significant in straight tubing than in coiled tubing. Shah and Zhou (2003) examined the influence of coiled tubing on the performance of DRP using four different polymer solutions (each of which was prepared in three different concentrations) in coiled tubing of 12.7, 38.1 and 60.3-mm internal diameters. The investigation revealed that drag reduction was higher in smaller tubing and as the tube size was increased, drag reduction was shifted to higher Reynolds number. The effect of polymer concentration was found to depend on the tubing diameter and flow rate – while the smaller tubing diameter favoured higher drag reduction with increase polymer concentration, higher flow resistance was observed at higher concentration and lower flow rate for the bigger tubing diameter. Also increase in curvature ratio was not favourable as it delayed and decreased the drag reduction. Among the four polymers used, xanthan gum and partially hydrolyzed polyacrylamide produced higher drag reduction than guar gum and hydroxyethyl cellulose and as compared with fluid flowing in straight tubing; there was a significant delay in transition from laminar to turbulent regime. The effect of coiled tubing curvature was further studied by Zhou et al. (2006) using 12.7-mm straight and coiled tubing sections with radius of curvatures of 0.01, 0.019, 0.031 and 0.076. It was discovered that drag reduction decreases with the increase in radius of curvatures in all the polymers used and tubing curvatures delayed the onset of drag reduction. This may be as a result of exhibition of significant friction pressure losses due to tubing small sizes and secondary flows caused by the curved flow geometry. Maximum drag reduction asymptote in coiled tubing was compared with the one obtained for straight pipe by Shah and Zhou (2009) using the previous flow loop. It was found that the two constants of the Virk›s asymptote correlation decrease with increase in the radius of curvatures.

Effect of Surface Roughness

The effect of surface roughness on drag reduction was reported by few investigators using different test section geometries (Virk, 1971, Petrie et al., 2003 and Karami and Mowla, 2012). Virk (1971) presented his results for different pipe roughness in terms of friction factor versus Reynolds number. He concluded that friction factor increases with the relative roughness. Karami and Mowla (2012) investigated the effect pipe roughness on pressure drop reduction in crude oil pipelines. Their finding revealed that drag reduction increases with the increase in pipe relative roughness. Petrie et al. (2003) conducted their experiments in rectangular channel using slot-injected and homogeneous polyethylene oxide cases over a range of flow conditions. The dimension of the rectangular cross section of the test section measured 508 mm in span by 114 mm in height by 762 mm in length. The results showed that even with fully rough surfaces, there was drag reduction of over 60% for both the injected and homogeneous polymer cases. The increase in surface roughness caused the drag reduction to decrease for both cases though the decrease was quicker in slot-injection method. Therefore, higher polymer concentration was needed to achieve the same level of drag reduction. This result is in disagreement with the one found by Karami and Mowla (2012) because the investigations were performed in external (channel) flows as compared to the internal (pipe) flow. It has been established that the primary problem with external flows with rough surfaces is the increase in the mixing which always occurs even at maximum drag reduction. The presence of roughness results in increased mixing of the polymer solutions and increased polymer degradation. This increase in the mixing rate eventually leads to the rapid decay of the near-wall polymer concentration and hence lowers the friction drag reduction. It is equally discovered that the presence of the roughness also led to an increased level of polymer degradation. These combined caused a significantly increased level of polymer requirement to achieve the equivalent levels of friction drag reduction over a long smooth surface (Ceccio et al., 2007).

Wavy surfaces have also been found to influence the drag reduction. Vlachogiannis and Hanratty (2004) carried out such an investigation in a rectangular channel flow where the top wall is a smooth surface while the bottom wall is a wavy surface with a wavelength of 5 mm and an amplitude of 0.25 mm. In the range of Reynolds numbers and concentration of polymer used, it is observed that drag reductions of the wavy surface were always higher than those of the smooth surface.

Effect of Molecular Weight

The influence of polymers molecular weight on drag reduction effectiveness has also been widely studied and there is general consensus that drag reduction increase with increases in the molecular weight of the polymers (Virk, 1975, Martin and Shapella, 2003 and Shanshool and Al-Qamaje, 2008). Martin and Shapella (2003) investigated the effects of molecular weight, and other parameters on drag reduction of oil-soluble polyisobutylene solutions in a 3.02-mm ID horizontal glass pipe. They found that the effectiveness of polyisobutylene as a DRP was enhanced by increasing the molecular weight, reaching up to 70% at the highest molecular weight. Shanshool and Al-Qamaje (2008) used the same polymer but with different molecular weights (2.9×10^6, 4.1×10^6 and 5.9×10^6) at 10–70 ppm concentrations in a gas oil turbulent pipe flow of 1.25-in. ID and found the same trend of drag reduction with molecular weight. In another study, it was discovered that the drag reduction increased with increase in molecular weight of the polymer and Reynolds number (Kim et al., 2009).

Effect of Chain Flexibility

The ability of flexible and rigid rod-like polymer to reduce drag was studied by a number of investigators (Sifferman and Greenkorn, 1981, Paschkewitz et al., 2005 and Japper-Jaafar et al., 2009). Sifferman and Greenkorn (1981) conducted their experiments using three different polymers (polyethylene oxide, sodium carboxymethyl cellulose and guar gum) with various concentrations ranging from 0.001 to 0.3 wt% injected in a 0.027-m ID galvanized pipe. Of the three polymers used, polyethylene oxide and guar gum resulted in drag reductions up to almost 80% at 0.3% concentration when the Reynolds number was in excess of 10^5. The sodium carboxymethyl cellulose was not as good as the other two, giving drag reduction of about 60%. The difference in the effectiveness of these polymers can be attributed to their flexibility. Polyethylene is well known to be more flexible than the remaining two and hence is expected to be more effective in drag reduction. In fact, these two polymers belong to semi-flexible group of polymers. However, being natural polymers with similar structures, the flexibilities of sodium carboxyl cellulose and guar gum are almost the same though the latter is a bit more flexible than the former. In addition, the drag reducing effectiveness of these two polysaccharides might be enhanced, especially guar gum to perform at the same level with polyethylene, by grafting flexible polyacrylamide (PAM) onto their rigid backbones. It has been established that copolymers with longer and fewer polyacrylamide branches are more drag reduction effective and more shear stable than those with shorter and more PAM branches (Singh et al., 1991). Paschkewitz et al. (2005) observed a drag reduction of about 10–15% and discovered that increasing the concentration of the polymer did not increase drag reduction but rather, the spatial development of the drag reduction was shifted further downstream. In addition to the observed drag reduction, three unique drag reduction regimes were developed, a non-Newtonian flow region close to the injector, followed by a nearly constant drag reduction region and a region of negligible drag reduction (see Fig. 6). It was also noticed that the decrease in the velocity resulted to a decrease in drag reduction.

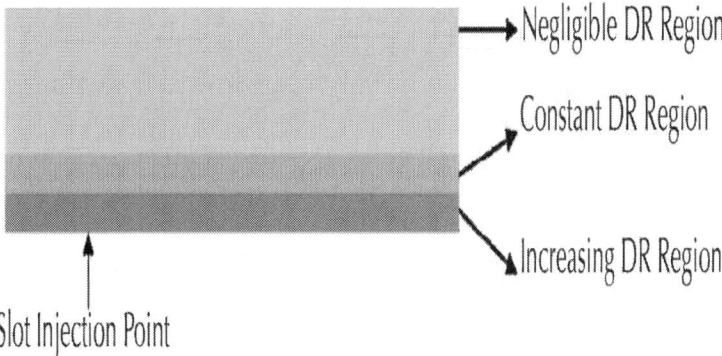

Negligible DR Region

Constant DR Region

Increasing DR Region

Slot Injection Point

Figure 6: Three distinct drag reduction regimes of turbulent boundary layer in a rectangular channel flow.

On the other hand, Japper-Jaafar et al. (2009) showed that the drag reduction obtained with the rigid rod-like polymer is lower than what was obtained from flexible polymer solutions. However, scleroglucan can still serve as effective drag reducing polymer. It was also noticed that the effectiveness of the polymer increase with increase in its concentration which contradict the earlier finding of Paschkewitz et al. (2005). Although there was no explanation given in regards to unusual result of earlier investigation, different flow geometries used by both studies might be responsible for the conflicting results. They made a suggestion that direct numerical simulations were required to fully understand the mechanism behind this type of polymer.

Effect of Polymer Structure and Composition

It was discovered that when shear-thinning polymer solutions are in transitional regime between laminar and turbulent flow, they exhibit significant departure from axisymmetry in contrast to the fully developed pipe flow of Newtonian fluids or both laminar and turbulent flows of these polymer solutions (Escudier and Presti, 1996, Escudier et al., 2005 and Peixinho et al., 2005). Escudier et al. (2009) carried out an experimental study to find an explanation for this asymmetric behaviour of aqueous solutions of polyacrylamide (300 and 1250 ppm) of molecular weight of 1.9×10^6 g/mol and xanthan gum (1500 ppm) of molecular weight of 5.1×10^6 g/mol by exploring the influence of imposing both upstream and downstream distortion on their flows. They concluded that the influence of the flow geometry at the upstream and downstream did not cause the asymmetric velocity fluctuation at the transitional regime of these flows.

Effect of pH and Temperature

Effects of temperature and pH on the drag reduction were studied by Interthal and Wilski (1985). Both polyethylene oxide (PEO) and partially hydrolyzed polyacrylamide (PAAM) of 30 ppm each were used in a pipe turbulent flow with 14 mm ID and 10^5 Reynolds number. The study showed that drag reduction of 80% achieved for PAAM was not affected by the temperature which was increased from 5 to 35 °C but drag reduction of PEO changed from 70 to 50% in the same temperature range. They ascribed this decrease to a reduction in solvation of the polymer molecules as the temperature increases. Meanwhile the drag reduction was found to depend highly on the pH of the solution. Alkaline polymer solution gave higher drag reduction than acidic counterpart with maximum of 80% at pH of 6.5. Conformation of polymer coil in the solution was assumed to be responsible for this change. According to Interthal and Wilski (1985), the carboxylic group in the acidic range tends not to be dissociated resulting to small and compact coil while in the alkaline range, dissociation is enhanced leading to opening up of the coil and thus a stronger interaction with the solvent will result. Another study of the effect of temperature used poly isobutylene in a 1.25-in. ID within the temperature range of 30–50 °C (Abid-Ali and Al-ausi, 2008). The result showed that the drag reduction increased with increase in temperature up to 45 °C where a maximum of 33% was achieved. However, it is not possible to ascertain whether the same trend of temperature and pH is true for all investigations as the published works on these parameters are very limited.

Effect of Solvent and Salt Content

Two types of polymer drag reduction behaviours (fan or Type A and ladder or Type B) have been found when the concentrations of salt contents in polymer solutions are varied. This behavioural change of polymer solution from Type A to Type B is related to the conformation of the polymer molecules, which in turn is dramatically influenced by the concentration of ions provided by the salt in the solution. For Type A drag reduction, which is typical of randomly coiled polymer solutions due to the high salt content, the onset of drag reduction occurs only when a certain wall shear stress is reached perhaps either because of uncoiling of the polymer chains in the extensional flow field or their entanglements reaching the size of turbulent eddies (Virk and Merill, 1969, Berman, 1978 and Virk and Wagger, 1989). On the other hand, Type B drag reduction is typical of extended polymer molecules due to low salt content and it exhibits asymptotic drag reduction immediately after transition from laminar to turbulent flow hence eliminating the normal laminar-turbulent transition region (Gasljevic et al., 2001). These two behaviours of polymer drag reduction are also related to diameter effect. For the same concentration of a polymer solution, it has been found that Type A is exhibited in larger pipes with lower wall shear stress while Type B is exhibited in smaller pipes with higher wall shear stress at the same Reynolds number (Liaw, 1969 and Liaw et al., 1971).

While Hoyt (1980) has noted that the significant decrease in the effectiveness of polyacrylamide polymers may have occurred due to the presence of rusts in pipes, drag reduction was found to depend on the type of solvent. This was more pronounced for oil-soluble polymers. Moreover, the presence of salt in the polymer solutions has been found to affect the magnitude of drag reduction (Interthal and Wilski, 1985, Martin and Shapella, 2003 and Henaut et al., 2009).

Interthal and Wilski (1985) added sodium chloride salt into PAAM solutions in the range of 0–3 wt% at 20 °C and found that the drag reduction was reduced from 80 to 77% at 0.5 wt% beyond which no decrease occurred. The effect of solvent solubility parameters on the drag reduction was investigated in oil-soluble polyisobutylene solutions with different solubility parameters. The maximum drag reduction achieved among the five various solvents was 68.8%, which corresponds to the solvent whose solubility parameter is equal to that of the polymer. This is attributed to the ease with which the polymer chains become more extended when the solubility parameters are equal thereby giving the chains the opportunity to have more and more contact with turbulent flow eddies and further reduce the drag.

The extent of drag reduction of water-soluble polymers can be increased by changing their side groups to give a solubility parameter closer to that of water (Martin and Shapella, 2003). On the effect of water content and waxes, four different oil-soluble polymers were used to study their capabilities of influencing drag reduction in three different crude oil flows. According to the findings by Henaut et al. (2009), the presence of these contaminants in the crude oils negatively affected the drag reducing ability of the drag reducing polymers.

Other DRP Studies in Single Phase Flow

Almost all the DRP related published works address separately drag reductions caused by either polymers or surfactants in turbulent flow. Hence, the synergistic effect of combined polymer and surfactant in drag reduction has received little or no attention. Mohsenipour and Pal (2013) investigated the effect of mixed nonionic polymer and cationic surfactant systems on drag reduction of turbulent pipe flow. Polyethylene oxide (PEO) was used as the non-ionic polymer at three different concentrations of 500, 1000, and 2000 ppm while octadecyltrimethylammonium chloride (OTAC) was used as the cationic surfactant at concentration levels of 1000 and 2500 ppm. In addition, sodium salicylate serving as a counter-ion for the surfactant was mixed with the surfactant in the molar ratio of 2:1. The results showed that there was appreciable interaction between the polymer and the surfactant based on the relative viscosity measurements. Higher drag reduction was obtained in the mixed polymer–surfactant flow system when compared with using either polymer or surfactant alone. However, the increase in drag reduction due to the addition of surfactant to polymer is more pronounced at low polymer concentration and high surfactant concentration. Therefore, this combination showed a strong synergistic effect in drag reduction of turbulent flow.

Mohsenipour et al. (2013) also conducted other similar experiments where they combined cationic surfactant with anionic polymer solutions. Employing deionized water and tap water as the flowing fluids, OTAC was used as the cationic surfactant while the polymer was changed to copolymer of acrylamide and sodium acrylate (PAM). The results revealed that the properties of the system were greatly increased when the surfactant was mixed with polymer solution. For instance, the critical micelle concentration (CMC) of the mixed surfactant–polymer system was found to be different from that of the surfactant alone while there was a significant decrease in the viscosity of the polymer. This decrease in the relative viscosity as a result of adding cationic OTAC was attributed to the change in the extension of polymeric chains thereby leading to the collapse of the chains. Therefore, the effectiveness of PAM in drag reduction was drastically reduced upon the addition of OTAC and this was quite noticeable at low concentrations of PAM. On the other hand, there is an interconnected network of polymer chains at high PAM concentrations and the effect of surfactant on drag reduction subsided due to the inaccessibility of PAM chains to surfactant molecules.

The nature of water (i.e. tap water or deionized water) used to prepare the polymer solution was found to influence the performance of the surfactant on drag reduction (Mohsenipour et al., 2013). Polymer chains became less water-soluble because of the neutralization by OTAC molecules thereby rendering the chains insoluble in deionized water and reduction in PAM drag reduction capability occurred. In the case of tap water, which is a poorer solvent than deionized water as result of the presence of cations, drag reduction ability was less affected when OTAC was mixed with PAM because the addition of OTAC to polymer solutions resulted in less shrinkage of the polymer chains.

Generally, all single-phase experiments with DRP were conducted using liquid as the fluid. In order to investigate the drag reduction effectiveness of DRP in gas phase flow, Gaard and Isaksen (2003) considered using various drag reducing additives in the flow of dense natural gas in 40-m long pipelines of different wall roughness. The main objective of the study was to find out the potential of the commercially available drag reducers for hydrocarbon liquids in reducing pressure drop and hence increase flow capacity of long distance dense gas pipelines. Although the names of the commercial DRPs tested were not mentioned, it was found that one of them brought about drag reduction efficiency of 19% in the pipe with the highest roughness at injection concentration of 150 ppm. On the other hand, drag reduction efficiency was obtained to be about 10% with the low roughness pipeline.

Industrial Use of DRPs for Crude Oil Transportation

Studies were conducted either in the laboratory or actual field trials to test the industrial use of DRPs. In the field, Lescarboura et al. (1971) tested a polymeric drag reducer in a single phase of crude oil flowing in a pipeline spanning 28 miles and with an ID of 8 in. They reported that for the normal flow-line velocity of 6 ft/s, drag reductions of 16, 21, and 25% were obtained at polymer concentrations of 300, 600, and 1000 ppm respectively. They also tested the DRP in 12-in ID, 32-miles pipeline, which showed a decrease in drag reduction by increasing pipe diameter. This is because at the same polymer concentrations of 300, 600, and 1000 ppm, drag reductions of 16, 16.2, and 16.9% were recorded. The major conclusions they draw from their tests are that drag reduction decreased with a decrease in flow velocity and increase with pipe diameter. It was also concluded that the tested polymer did not degrade or lose its effectiveness as a result of turbulence or shear in the flow lines.

The first industrial application of DRP was in 1979 when 10 ppm oil-soluble polymer was employed to increase flow rates of the 1300-km Tans-Alaska oil pipeline system, which is 1.2-m in diameter. Drag reduction of 50% was achieved thereby eliminating the need for installing additional two pumping stations to raise the throughput from 1.45 to 2.1 million barrel per day (MBPD) (Burger et al., 1982 and Jubran et al., 2005). Apart from this, Interthal and Wilski (1985) has also conducted industrial scale tests. Their results showed that drag reductions of 75, 65, 64 and 65% were achieved through the 75, 300, 450 and 750-mm ID pipelines respectively. Again, there was no consistent trend of drag reductions with pipe diameter even at industrial level.

DRAG REDUCTIONS IN MULTI-PHASE FLOW EXPERIMENTS

Multiphase flow occurs when two or more fluid phases flow simultaneously through a conduit. Phases tend to separate and flow at different velocities due to different densities resulting in variation in the physical distribution of the phases in the conduit. This physical distribution of phases results in different flow patterns governed by forces such as buoyancy, turbulence, inertia and surface tension. To reduce the turbulence and drag by prolonging the laminar flow or delaying the onset of turbulent flow to a higher Reynolds numbers, DRPs are injected into the flow.

Unlike the single phase flow, flow pattern is one of the most important features of multiphase flows in pipes as it determined the pressure drop involved and how easy it is to achieve separation of the phases. Each type of flow pattern depends on the relative velocities of the phases in the multiphase. Although flow of gas–liquid flow is presented in Fig. 7, stratified, slug and annular flows are predominant in oil and gas industries. Stratified flow is a separated layer of gas and liquid which may contain wave at the interface. Slug flow is characterized by a region of long bubble of gas and region of intermittent liquid slug containing dispersed bubble of gas. Annular flow on the other hand, is a form of flow with a continuous gas core often conveying trapped liquid droplets bounded by a thin film on the tube wall (de Schepper et al., 2008).

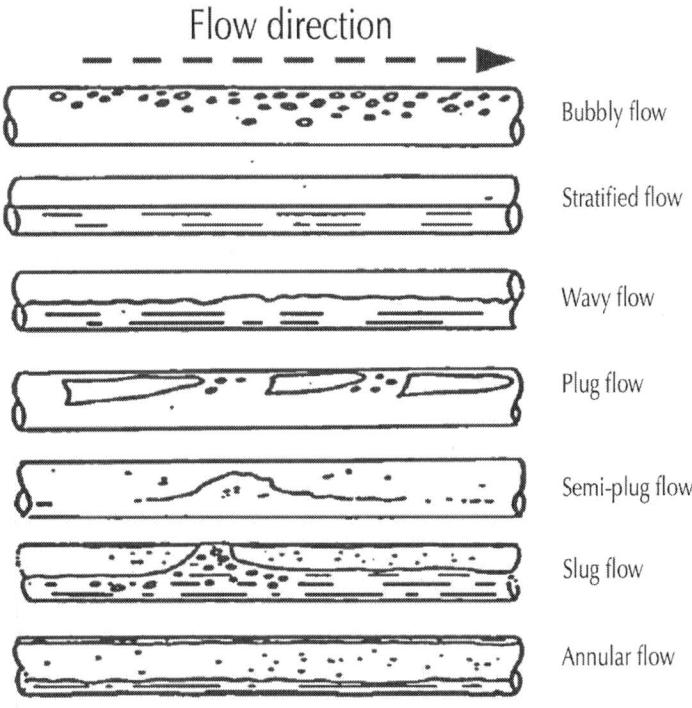

Figure 7: Two-phase gas–liquid flow patterns in horizontal flow.

Presently, majority of published results on DRPs in multiphase flow, most of which on two-phase gas–liquid flow, have come from few research studies that cover a variety of experimental set-ups, pipe diameters, injection techniques, DRA chemicals, concentrations, flow rates, and flow regimes (Langsholt, 2012). The following sections present a review of these experimental efforts involving the injection of DRPs to reduce turbulence and drag in multiphase flows.

Two-Phase Gas–Liquid Flow Experiments with DRPs

Two-phase co-current gas–liquid flows are frequently encountered in the industries and particularly, they are of great commercial importance in the natural gas and petroleum industries. For example, offshore production has made the transportation of gas–liquid two-phase flow over long distances before separation inevitable. Depending on the flow rates of the gas and liquid, one can broadly classify the flow patterns of gas–liquid flows into stratified, bubble, annular and slug flows. The reduction of the pressure gradient, from the economic point of view, is of practical importance as it can reduce the pumping energy of the fluid in long pipelines and quantities of gas with associated liquid can be transported without building a new pipeline or increasing pump horsepower. Because of the ability of certain polymers to influence the pressure drop and possibly the flow patterns, the study of the drag reduction polymers will not only help in the design of process equipment but will also assist in understanding the potentiality of the polymers in lowering the pressure loss during the transportation of the two-phase gas–liquid over long distances in the oil and gas industries.

Drag Reduction

Several experiments have been carried out on the effect(s) of DRPs on gas–liquid flow in pipes. Oliver and Young (1968) was one of the first works that investigated drag reduction on a mixture of polyethylene oxide solution and air. They found that in annular flow, wave formation was damped, while in slug flow, the liquid showed less circulation.

Greskovich and Shrier (1971) who initiated studies on the use of DRP in multiphase systems investigated the effect of un-named polymer DRA in a 0.038-m diameter horizontal pipe. In fact, they were the first to use the abbreviation DRP in the study of multiphase flow. They reported that about 40–50% of drag reduction could be achieved in air–water slug flow and that frictional loss reduction as opposed to change in acceleration within the slug is responsible for this drag reduction. In a similar work by Rosehart et al. (1972)who used polyacrylamide in slug flow in a 0.0254-m diameter horizontal pipe, it was generally concluded that there was no change in slug characteristics like translational velocity. However, drag reduction was most noticed at high slug velocity which can be attributed to skin frictional and energy losses. It was also observed that there was a drag reduction asymptote beyond which no additional drag reduction was obtained with the increase in the DRP concentration.

Scott and Rhodes (1972) introduced polyacrylamide commercially known as Polyhall295 into a slug mixture of air and water flowing through 2.5 cm ID pipeline with constant liquid Reynolds number and various gas Reynolds numbers. Based on their findings, they concluded that drag reduction was higher in two-phase flows than in single-phase flows at the same superficial velocities and a maximum of 33% drag reduction was found to occur at a polymer concentration of 68 ppm. In addition to what Rosehart et al. (1972) obtained, it was noticed that the observed pressure drop was always lower than that predicted by the Lockhart–Martinelli correlation. However, the maximum drag reduction followed the Arunnalchum modification of Virk's drag reduction asymptote (Otten and Fayed, 1976).

Sylvester and Brill (1976) carried out investigation into the effect of DRPs on annular air–water flow in a 6.1-m long horizontal pipe with 1.27-cm ID using 100 ppm polyethylene oxide solution. The results showed that drag reduction ranging from 0 to 37% were achieved. This work was extended in 1980 when they investigated the effect of a DRP on an annular-mist flow of natural gas–hexane (Sylvester et al., 1980). Their results showed that in the annular-mist flow regime, drag reduction increased with decreasing gas flow rate for a given liquid flow rate and also increased with increasing liquid flow rate at a fixed gas flow rate. They achieved a maximum drag reduction of 37% and observed that drag reduction decreased as the liquid–gas ratio tends towards zero (Sylvester et al., 1980).

Kang and Jepson (1999) carried out an experimental study of gas–oil flow in a 10-cm ID, 36 m long multiphase inclinable

Plexiglas flow loop using superficial liquid velocities between 0.5 and 3 m/s and superficial gas velocities between 2 and 10 m/s in 2° inclined pipes. The un-named DRP concentrations of 10 ppm and 50 ppm were injected into the oil which has a viscosity of 2.5 cP oil at 25 °C. It was observed that the DRP was effective in reducing the pressure gradient at all liquid and gas velocities. At a superficial liquid velocity of 1 m/s and gas velocities of 2 and 6 m/s, the corresponding drag reductions were 31% and 40% respectively. In a similar experiment conducted with the same fluids and flow loop but in a horizontal configuration, there was more significant pressure drop reduction for all superficial liquid and gas velocities in both slug and annular flow. Drag reduction values were 82% for slug flow and 47% for annular flow (Kang and Jepson, 2000).

Al-Sarkhi and Hanratty (2001a) investigated how a co-polymer of polyacrylamide and sodium acrylate (Percol 727) affected an annular flow of air and water in a horizontal pipe having an internal diameter of 0.0953 m and length of 23 m. 10–15 ppm of the polymer was injected, with gas velocities of 30, 36, and 43 m/s and the liquid velocities ranging from 0.034 to 0.147 m/s. Results of their experiments showed that drag reduction of about 48% was achieved at low concentrations of polymers in water. At large polymer concentrations, annular flow regime changed to stratified flow regime. Effect of DRPs on interfacial drag of pseudo-slugs and their transition to fully slug flow was studied. Pseudo-slugs are large amplitude roll waves that cover the interface of air and water flowing in a stratified configuration as a result of the increase in superficial liquid velocity at a fixed superficial air velocity. In the experiment conducted in a horizontal pipe of 2.54-cm ID and 18.3-m long, it was observed that waves with small wavelength were damped causing a significant decrease in the interfacial stress, which increased the liquid holdup. It was also noted that pressure drop can be increased or decreased depending on the net effect of stresses caused by the increased liquid hold-up. An increase in polymer concentration increased liquid hold-up and velocity, which was accompanied by a decrease in the roughness at the interface. The increase in velocity increases wall and interface resisting stresses while the decrease in roughness decreases interfacial stress (Soleimani et al., 2002). In a brief communication by Baik and Hanratty (2003), wall drag reductions of 42% were realized for USG = 1.5 m/s, USL = 0.15 m/s and 30% for USG = 5 m/s, USL = 0.14 m/s. They used partially hydrolyzed polyacrylamide (HPAM, Magnafloc 1011) in a horizontal Plexiglas pipe that had a diameter of 9.53 cm and a length of 23 m for air and water combined in a tee-section at the entry point.

Fernandes et al. (2004) studied the effect of poly alpha-olefins on a mixture of methane and condensate flowing through a horizontal pipe under high pressure. They concluded that drag reduction in an annular flow is mainly due to flow pattern modification. Al-Sarkhi and Abu-Nada (2005) investigated the effectiveness of Magnafloc 1011, which is a copolymer of polyacrylamide and sodium-acrylate, in the drag reduction of an annular flow of air–water mixture in a horizontal pipe. They achieved a maximum DR of 71% with 100 ppm of polymer injected at a pipe inclination of 1.28°. They concluded that the effectiveness of DRPs may be dependent on the gas and liquid flow rates. Mowla and Naderi (2006) investigated the effect of poly isobutylene on slug flow of air and crude oil using a flow loop. Their results showed drag reduction of about 40% under some experimental conditions. Also they concluded that the applied DRP is more effective in rough pipes than in smooth pipes and that drag reduction is higher in the 0.0127-m ID pipe than in the 0.0254-m ID pipe.

Parimal et al. (2008) examined the effect of both oil soluble and water soluble DRPs on average pressure drop and slug characteristics of air and water flow in a 36-m long pipe of 10-cm ID. They observed that the addition of the DRPs resulted in significant decrease in pressure drop and turbulence. They also noted that the injection of the oil soluble DRP may have a negative impact on the flow especially if the flow is in the dispersed flow regime because it induced emulsion formation.

Xu et al. (2009) used new approach to study the phenomenon of drag reduction in gas–liquid system in horizontal Perspex 50-

mm ID, 30-m-long pipeline. This was conducted by injecting gas into carboxymethyl cellulose solution in stratified and slug flow regimes. The liquid hold up was measured with two rapid closing valves and the flow patterns were observed by a high-speed video camera. A total of 180 experiments were conducted by changing superficial liquid velocity from 0 to 1.42 m/s, superficial gas velocity from 0 to 2.59 m/s and input liquid phase cuts from 0 to 100%. It was discovered that gas injection alone caused drag reduction even before the addition of the polymer. In summary, the results showed that for turbulent gas–laminar liquid stratified flow, the drag reduction by gas injection is more effective for Newtonian fluid than that for the shear-shinning fluid at high dimensionless liquid height. Also the drag reduction was noticed over the large range of the liquid holdup at low flow behaviour index.

Recently, the effect of polymer mixing on the performance of DRP was investigated in annular horizontal two-phase flow (Al-Sarkhi, 2012). The experimental setup consists of a pipeline with ID of 95.3 mm and length of 23 m using Plexiglas material to allow visual observation. The polymer master solutions (Percol 727) with concentrations ranging from 250 to 1500 ppm were injected into the flow loop in two ways: firstly at the entrance injection point 0.6 m before the air–water mixing tee section and secondly at 3.7 m downstream in tri-channels (the channels are vertical and at ±15° from the vertical) after the mixing tee. It was observed that the mechanism of interaction of the polymer aggregates with the disturbance waves in the flow was influenced by the method of injecting the polymers and the polymer concentration in the master solution. It was concluded that polymer aggregates must be active and not disintegrated to be able to dampen the turbulence which was achieved when the injection was downstream the mixing tee. In addition, the onset of drag reduction was lower at higher concentration of the master solution. On the other hand, the onset of drag reduction occurred at lower concentration of downstream injection than entrance injection (Al-Sarkhi, 2012).

Effect on Flow Patterns

The addition of DRPs on gas–liquid flow does not only reduce the pressure drop but also change the flow patterns. It has been experimentally found over a wide range of conditions that the presence of a DRP in annular flow causes damping of wave formation and results in changing the flow pattern from the annular flow to either another annular flow with a much smoother gas–liquid interface, very low entrainment and more stratified appearance or to a fully stratified flow with virtually no entrainment (Oliver and Young, 1968, Al-Sarkhi and Hanratty, 2001a, Al-Sarkhi and Hanratty, 2001b, Al-Sarkhi and Soleimani, 2004 and Al-Sarkhi and Abu-Nada, 2005). This change in flow pattern may be due to the elimination of the disturbance waves and the reduction in the amount of entrained droplets by the added DRP (Al-Sarkhi and Hanratty, 2001a and Al-Sarkhi and Hanratty, 2001b). Further investigation by Al-Sarkhi and Soleimani (2004)showed that annular flow may change to stratified flow, wavy stratified flow may change to smooth stratified flow, and slug flow may break up into a more unstable form of slug flow or pseudo slug flow (Fig. 8 andTable 1). According to the review carried out by Al-Sarkhi (2010), DRPs cause much less circulation of liquid in slug flow and dampen wave formation in annular flow which results to a smoother liquid film.

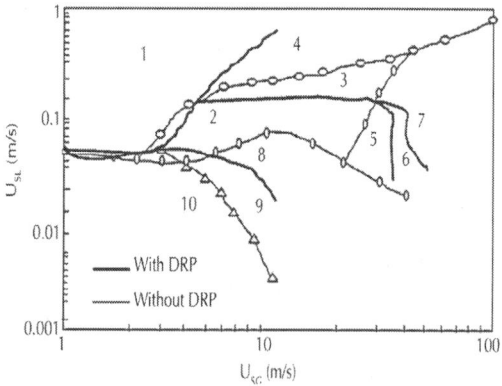

Figure 8: Flow pattern map of air–water flow in 2.54 cm horizontal pipeline with co-polymer of polyacrylamide and sodium acrylate of concentration 0 and 50 ppm (Al-Sarkhi and Soleimani, 2004).

Table 1: The names of the flow patterns represented in Fig. 8 (Al-Sarkhi and Soleimani, 2004)

Number	Without polymer	With polymer
1	Slug	Slug
2	Pseudo slug	Stratified
3	Pseudo slug	Pseudo slug
4	Slug	Pseudo slug
5	Annular	Stratified
6	Annular	Stratified-annular
7	Annular	Annular
8	Wavy stratified	Wavy stratified
9	Wavy stratified	Smooth stratified
10	Smooth stratified	Smooth stratified

Baik and Hanratty (2003) also studied the effect of a DRP on a stratified gas–liquid flow. They measured the interface height using a conductivity probe with and without DRP. Their results showed that the DRP profoundly reduced the interfacial wave structure and the reduction in wave activity leads to a reduction in the interfacial friction factor, which increased the average thickness of liquid layer. Although the addition of polymers delayed the transition to slug flow, no effect of polymers on the critical superficial liquid velocity for the transition to slugging was observed for high superficial gas velocity.

Like gas–liquid stratified flows, slug flow is also very complex and so is the effect of DRPs on this flow pattern. Kang and Jepson (2000) found that at superficial liquid and gas velocities of 0.5 and 2 m/s, gas–oil slug flow was changed to stratified flow with the addition of 50 ppm DRP leading to a decrease in the slug frequency to zero. The effective liquid film height was found to decrease in the presence of DRP and the slug frequency at higher liquid and gas velocities decreased significantly. Other results on the effects of DRPs in slug flow are reported by Rosehart et al. (1972), Otten and

Fayed (1976), Saether et al. (1989), Al-Sarkhi and Soleimani (2004) and Al-Sarkhi (2010). In general, it is believed that the turbulence in the slugs is damped by addition of DRP and this leads to an increase in shedding rate of the slug and therefore its stability. Soleimani et al. (2002) studied the effect of co-polymer of polyacrylamide and sodium acrylate on a pseudo-slug flow of air and water in a horizontal pipe 0.0254-m ID. Damping of waves and increase in liquid hold-up were observed whereas transition to fully developed slug was delayed at superficial gas velocities greater than 4 m/s as it occurs at larger liquid holdups. They then ascribed this observation to dampening of the turbulence in slugs.

Two-Phase Liquid–Liquid Flow Experiments with DRPs

The simultaneous flow of two immiscible liquid in pipes is a common phenomenon in chemical and petrochemical industries. The flow is usually stratified at low velocities, but as flow-rates increase, transition from stratified to non-stratified flow patterns occur, and eventually transform to disperse flows as velocities are further increased. In the disperse regime, significant pressure drop is experienced and phase inversion phenomenon, which is usually associated with high pressure drop in flow, also occurs (Ioannou et al., 2005). This leads to higher power consumption and ultimately high cost of production. Injection of enhancers, known as DRPs into the flow can be used as an alternative solution to decrease this high pressure drop. Also, as oil–water mixtures are difficult to separate at the end of the pipeline, these polymers will not only reduce the pressure drop but also preserve the stratified pattern for a wider range of conditions which will facilitate oil–water separation (Al-Sarkhi, 2010 and Yusuf et al., 2012).

Drag Reduction

Studies involving DRPs in liquid–liquid two-phase flow is actually very limited as summarized in Table 2. However, the first documented work on this type of flow was carried out by Al-Wahaibi et al. (2007) where they reported the effects of two concentrations of a co-polymer (Magnafloc 1011) injected into a horizontal flow of oil and water. The investigation was performed in 14-mm ID acrylic pipe using oil density of 828 kg m^{-3} and viscosity of 5.5 mPa s, and 1000 ppm polymer master solutions to provide 20 and 50 ppm into the water. Their finding showed that there was reduction in the pressure drop that became more pronounced as the water velocity increased particularly when there was a change in the flow pattern and due to injection of DRP. There was also a maximum drag reduction of about 50% when the polymer was introduced into annular flow. Similar results were obtained in the work of Al-Yaari (2008) using kerosene as the oil phase.

Table 2: Experimental studies of the effect of DRPs on liquid–liquid flows

Author's name(s)/year	Fluid(s) type/ properties	Pipe material(s)/ geometry	Polymer used/quan- tity (ppm)	Parameter (s) inves- tigated/ studied	Measurement technique (s)	Max. DR (%)
Al-Wahaibi et al. (2007)	Oil–water/ μo = 5.5 mPa s,\ ρo = 828 kg m^{-3}	Acrylic/ horizontal ID = 14 mm L = 3.5 m	Magnafloc 1011[a]/ 20, 50 ppm	Effect of DRP on the flow pat- terns, hold- up, Pressure drop	Visual observation, Differential manometer	50
Al-Yaari et al. (2009)	SAFRA D60– water/μo = 1.57 mPa s, ρo = 780 kg m^{-3}	Acrylic/ horizontal ID = 25.4 mm L = 10 m	Magnafloc 1011[a]/ 10–15 ppm	Effect of DRP on flow, pres- sure drop; Effect of DRP Conc., MW; Salt cont.	Visual observation, Pressure transducer	65
Omer and Pal (2010)	3 water-in-oil emulsions/ μo = 2.5, 6, 5.4 mPa s, ρo = 753, 785, 816 kg m^{-3}	4 stainless steel pipes/ horizontal ID = 8.9, 12.6, 15.8, 26.5 mm and 1 PVC pipe/ horizontal ID = 23.7 mm	Polyethylene oxide and carboxy- methyl cel- lulose/ 0–1 wt% each	Effect of DRPs on dispersed aqueous phase of the water-in-oil emulsion	Pressure transducers	Not stated
Al-Yaari et al. (2012)	SAFRA D60– water/ μo = 1.57 mPa s, ρo = 780 kg m^{-3}	Plexiglas/ horizontal ID = 25.4 mm L = 10 m	Magnafloc 1011[a]/ 50 ppm	Effect of PDRA on water hold- up	Conductiv- ity probe, Pressure transducer	42

Langsholt (2012)	ExxsolD80-water/ μo = 1.8 mPa s, ρo = 22.5 kg m^{-3}	Near horizontal, ID = 100 mmL = 25 m	Acrylic copolymer[c], poly α-olefin[d]/ 120[c], 80[d] ppm	Effect of DRP on the hold-up, Pressure drop, Diameter scaling	Beam gamma densimeters, Pressure transducer	30[c], 10[d]
Yusuf et al. (2012)	Mineral oil–water/μo = 12 mPa s, ρo = 875 kg m^{-3}	Acrylic/ horizontal ID = 25.4 mm L = 8 m	Magnafloc 1035[b]/ 2–10 ppm	Effect of DRP on flow patterns, drag reduction	Visual observation, Dywer 490 DDM[e]	60
Al-Wahaibi et al. (2013)	Mineral oil–water/ μo = 12 mPa s, ρo = 875 kg m^{-3}	2 Acrylic pipes/ horizontal ID = 19, 25.4 mm L = 8 m each	Magnafloc 1035[b]/ 2–30 ppm	Effect of pipe diameter on the performance of DRP	High-speed camera, Visual observation, Dywer 490 DDM[e]	45 for 19 mm, 60 for 25.4 mm

[a]Co-polymer of polyacrylamide and sodium-acrylate.

[b]Highly anionic co-polymer of 40:60 wt/wt NaAMPS/acrylamide.

[c]Water-soluble.

[d]Oil-soluble.

[e]Digital differential manometer.

Al-Yaari et al. (2009) injected 10–15 ppm from 1000 ppm polymer master solution using polyethylene oxide of molecular weights of 3×10^5, 4×10^6 and 8×10^6 into 10-m long acrylic (to allow visual observation) horizontal pipe with 25.4-mm ID. It was found that pressure drop reduction was significant and depends on water fraction, mixture velocity, concentration and molecular weight of the DRP. The presence of salt content in the water phase negatively affected the effectiveness of the DRP.

Yusuf et al. (2012) investigated the effect of DRP on pressure drop using higher viscosity oil in 25.4-mm ID, 8-m long acrylic pipe. Drag reduction increased with the polymer concentration from as low as 2 ppm up to 10 ppm beyond which no further increase in the drag reduction was observed. The same trend was also observed with the increase in water superficial velocity where the maximum drag reduction was achieved at 1.3 m/s. Increase in oil superficial velocity at water superficial velocity greater than 1.3 m/s decreased the drag reduction while no obvious trend in the drag reduction was noticed below water superficial velocity of 1.3 m/s. They noticed that the concentration of the polymer master solution did not affect the drag reduction and pipe length had little effect. Langsholt (2012) investigated the effect of adding DRP to either oil or water phase on the pressure drop, holdup and flow patterns. They concluded that drag reduction increased with increase in inlet fraction of the phase in which DRP was injected. In dispersed flow, drag reduction occurred even when the liquid carrying DRP was not the continuous phase.

Effect on Flow Patterns

The change from stratified flow pattern at low velocities to non-stratified flow pattern at high velocities (Fig. 9) causes huge pressure drop and phase inversion, which leads to higher pumping capacity and difficulty in the separation of the phases.

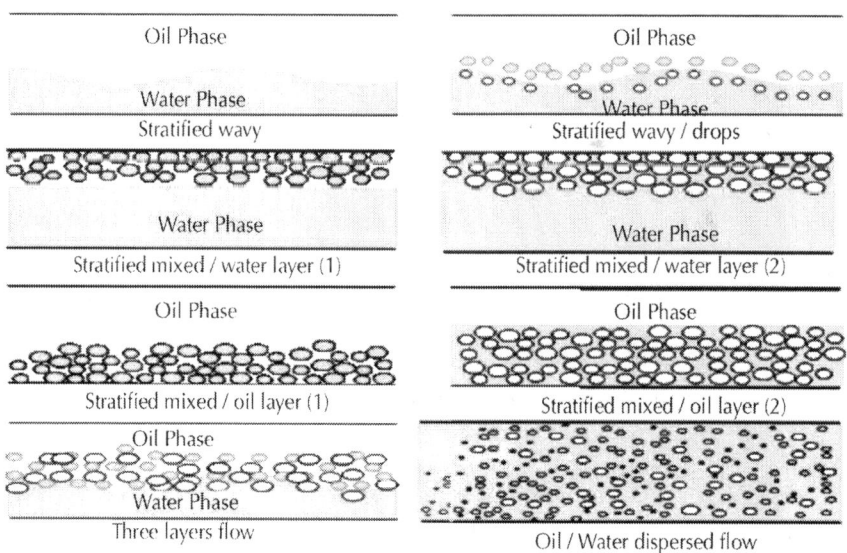

Figure 9: Observed oil–water flow patterns in a horizontal 0.0254 m pipe (Al-Yaari et al., 2009).

Therefore, a delay in flow pattern transition from stratified to non-stratified is highly needed to reduce energy consumption and enhance effective separation in two-phase liquid–liquid flow system. Like in gas–liquid flow, the presence of DRPs in liquid–liquid flow can influence the phase distribution of the flow thereby making the separation of the phases easier. For instance, the work of Al-Wahaibi et al. (2007) showed that on the addition of DRP to oil–water flow, stratified flow region was extended, transition to slug flow was delayed and annular flow was eliminated (Fig. 10). In the range of flow condition investigated, there was dampening of interfacial waves, which resulted in changing the annular flow to either stratified flow or dual continuous flow while slug flow changed mostly to stratified flow. Al-Yaari et al. (2009) observed similar results and the six flow patterns exhibited by oil–water flow without DRP were reduced to five after polymer was added. Again there was an extension of stratified and three layer flows in a wide range of mixture velocity and water fraction while a reduction in the dispersed flow occurred at high water fraction. They attributed the change from dispersed to stratified flow to the increase in droplet coalescence and reduction in turbulent mixing when DRP was added (Fig. 11).

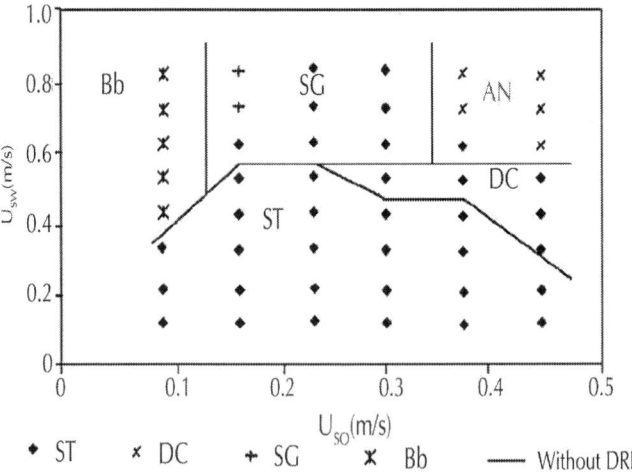

Figure 10: Comparison of flow pattern boundaries of oil–water flow without DRP with oil–water flow with 50 ppm DRP in the 14 mm ID acrylic horizontal pipe. The flow pattern (AN) which does not appear when 50

ppm DRP is added is written on the map in grey colour. Flow patterns: ST (stratified smooth or wavy), DC (dual continuous), SG (slug), Bb (bubble) and AN (annular) (Al-Wahaibi et al., 2007).

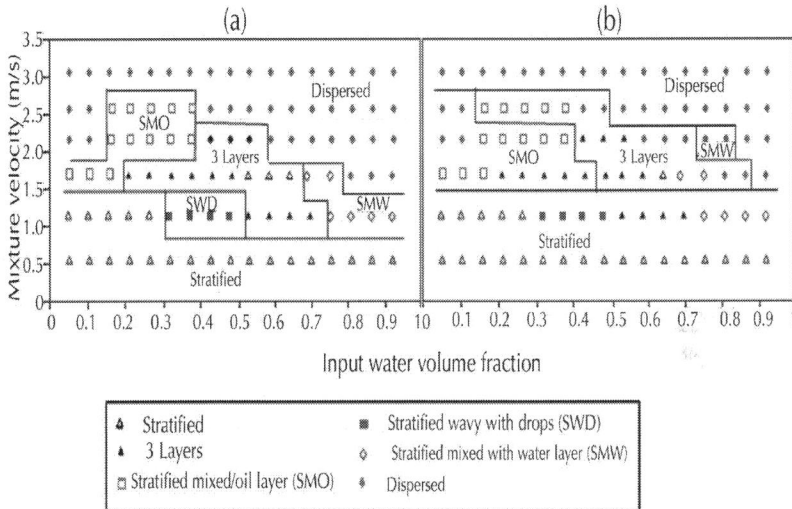

Figure 11: Flow pattern map of oil–water flow: (a) without DRP, (b) with 50 ppm of polyacrylamide solution (Al-Yaari et al., 2009).

The extension of the work of Al-Yaari et al. (2009) was carried out using the same polymer and experimental flow loop to study the effect of DRP on the water hold-up of the oil–water flow. Maintaining the mixture velocity at 1 m/s by simultaneously varying both oil and water superficial velocities from 0.1 to 0.9 m/s, the presence of DRP changed stratified wavy flow to smooth stratified at USW ≤ 0.2 m/s, stratified wavy flow with droplets at the interface to stratified wavy between 0.35 ≤ USW ≤ 0.5 m/s, three layer flow to stratified wavy and stratified wavy flow with droplet at the interface between 0.55 ≤ USW ≤ 0.7 m/s and stratified wavy with mixed layer at the top to stratified wavy flow with droplets at the interface between 0.75 ≤ USW ≤ 0.9 m/s. The water hold-up was found to be affected by DRP which dampened the high amplitude waves and reduced the turbulence leading to decrease in droplets formation. BelowUSW = 0.5 m/s, the water hold-up showed an increase while it decreased above USW = 0.5 m/s. It was concluded that the elongational viscosity of the polymer solution could be also responsible for the drag reduction in oil–water flow. Shear thinning behaviour of the polymer as well as injection method could play a role in wave growth and structure (Al-Yaari et al., 2012). The results of experimental study by Yusuf et al. (2012) did not show any deviation from previous observations as stratified and dual continuous flows extended to higher superficial oil velocities while annular flow also changed to dual continuous flow.

Two-Phase Solid–Liquid Flow Experiments with DRPs

There are few reported studies with respect to the effect of DRPs on solid–liquid flow. It has been found that solid particles alone, like nylon and cotton, suspended in a liquid flow can cause drag reduction (Radin et al., 1975). They concluded that only fibrous additives can cause drag reduction in liquid flow. Recently, cocoa nut fibre waste was added to water flowing in pipeline to act as drag reducing agent where four testing section lengths of 0.5, 10.0, 1.5 and 2.0-m were used. It was observed that drag reduction increased with testing section length reaching a maximum of 55.6% at 2.0-m testing section. In addition, drag reduction was found to depend on concentration of the fibre and flow rate of the fluid (Marmy et al., 2012).

Sifferman and Greenkorn (1981) conducted an experiment to study the effect of three different polymers on sand–water mixture. Of the three polymers used, guar gum and polyethylene oxide (POLYOX) were found to be much more efficient than sodium carboxymethyl cellulose (CMC). Also, drag reductions in the range of 95–98% were observed for the sand–polymer solution system and interestingly, the contribution of the sand to this drag reduction was more than the polymer itself. Therefore it was suggested that the mechanism of drag reduction may be additive (i.e. both the polymer and the sand have their separate drag reducing ability with the same mechanism). Effect of silica sand size and concentration on drag reduction was also investigated in a sand–polymer solution. The polymer name is Xanvis L and the solution flowed in a smooth-surface PVC vertical pipe of 2.54-cm ID and 0.259-ft long. The sand particles were 0.150–0.180, 0.425–0.60 and 2.00–2.36 mm in sizes while the concentration of sand ranges from 2.2 to 9.3%. Even though the pressure drop of sand–polymer solution at low Reynolds number was higher than sand–water mixture without the polymer, the results showed that the presence of polymer significantly reduced the pressure gradient at high Reynolds numbers, causing a drag reduction up to 60%. In contrast to the general perception, it was found that pressure drop decreased with increase in particle size of the silica sand. However, there was no explanation given to support this contrasting finding (Chung et al., 2007).

Three Phase Flow Experiments with DRPs

Oil–water–gas mixtures flow in pipelines is a common occurrence in the petroleum production mostly as a result of the injection of gas into the well to enhance the recovery of the oil. Therefore, reduced pressure drop of this multiple phase flow for long distances with stratification and hence easy separation of the phases are required. Although the use of DRPs can help in achieving this aim, very little documented studies can be found in literature. Sifferman and Greenkorn (1981) investigated three-phase oil–polymer solution–sand mixture and obtained similar result with the two-phase solid–liquid system. Prominent among the few experimental studies on the effect of DRPs on gas–liquid–liquid is the study of Kang et al. (1998). They carried out experiments on three-phase CO_2–oil–water flows in a 100-mm ID, 18-m long pipe. They found that the drag reduction obtained for a given concentration of DRP depends on the water cut and flow patterns while obtaining up to 81% drag reduction for stratified flow and 35% for annular flow. They concluded that the addition of DRPs delayed the stratified-slug transition to higher superficial liquid velocities, but did not alter the superficial gas velocity.

Another elaborate investigation was carried out by Langsholt (2012) where he injected two partially degraded polymers (water-soluble acrylic copolymer and oil-soluble poly α-olefin) into multiphase oil–water–gas flowing in a near-horizontal pipe of 100-mm ID and a length of 25 m. The stratified three-phase flow pattern was at 1° downward inclination and the results showed that drag reduction depends on the liquid into which the polymer was injected and increases with increase in inlet fraction of the polymer-carrying liquid. Lowering the gas velocity favoured the drag reduction and liquid hold-up especially for the case where the polymer was injected into water. For the slug three-phase flow which was oriented at 1° upward inclination, there was influence of the DRPs regardless of which liquid carried the polymer even though water-soluble DRP was only marginally better than oil-soluble one. Also, drag reduction increased with inlet liquid fraction carrying the DRP while the superficial velocities of the phases did affect the drag reduction. Similar results were obtained for annular three-phase flow in terms of the inlet fraction of the liquid carrying the DRP while more than 50% drag reduction was achieved at the highest superficial gas velocity.

Parameters Influencing the Performance of DRPs in Multiphase Flow

Effect of Channel Size and Roughness

One of the factors that influence drag reduction of DRPs is the pipe diameter (or the friction factor which is proportional to pipe ID) as has been shown by Al-Sarkhi and Hanratty, 2001a and Al-Sarkhi and Hanratty, 2001b. They conducted flow experiments using a DRP concentration of 30 ppm through pipe diameters of 0.0254 m and 0.0953 m. They found that maximum drag reduction increased with a decrease in pipe diameter in air–water annular flow. However, larger concentrations of DRP were required to obtain maximum drag reduction in the smaller pipe compared to the bigger one. In addition, Mowla and Naderi (2006) investigated into the effect of poly isobutylene on slug flow of air and crude oil using a flow loop comprising different pipe sections, one of which was smooth polycarbonate horizontal pipe of length 10.3 m and 0.0254-m ID and the other two pipes were rough galvanized iron of equal lengths of 8.8 m and IDs of 0.0254 and 0.0127 m. They found that drag reduction also increased with decrease in pipe diameters while it increased with increase in the pipe roughness. Moreover Al-Wahaibi et al. (2013), found that a decrease in pipe diameter tend to extend the stratified region to higher superficial oil velocity (Fig. 12).

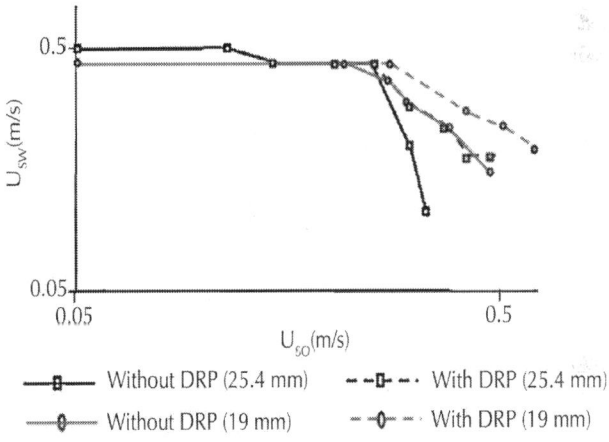

Figure 12: Comparison of stratified region in 19 and 25.4 mm pipes, with and without polymer (Al-Wahaibi et al., 2013).

Effect of Inclination Angle

Pipe inclination angle can also influence the achievable drag reduction in multiphase flow. For instance,Kang and Jepson, 1999 and Kang and Jepson, 2000 conducted experimental study on gas–liquid flow in a 10-cm ID pipe oriented at 2° and horizontal directions using an un-named DRP. The results showed that there was more drag reduction in the horizontal configuration than in the inclined one. Similarly, Al-Sarkhi et al. (2006) conducted an air–water flow experiment to determine the effect of pipe inclination and the performance of DRPs. Results of their experiments showed that at an inclination angle of 1.28° and polymer concentration of 100 ppm a maximum drag reduction of 71% was obtained while a higher pipe inclination of 2.4° resulted in the lowest drag reduction. They concluded that in general, the higher the inclination angle, the lower the drag reduction. Apart from its effect on drag reduction, pipe inclination also influences the flow patterns of air–water flow (Table 3) (Al-Sarkhi et al., 2006).

Table 3: Flow patterns of air–water annular flow with and without 100 ppm Magnafloc 1101 polymer (Al-Sarkhi et al., 2006)

U_{SG}(m/s)	U_{SG}(m/s)	Without DRP all angles	With 100 ppm ($\vartheta = 0°$)	With 100 ppm ($\vartheta = 1.28°$)	With 100 ppm ($\vartheta = 2.4°$)
38	0.10	Annular	Annular-clear	Annular-clear	Annular-clear
38	0.08	Annular	Annular clear	Annular-clear	Annular-clear
38	0.07	Annular	Annular-stratified	Annular-clear	Annular-clear
38	0.05	Annular	Annular-stratified	Annular-clear	Annular-clear
38	0.04	Annular	Annular-stratified	Annular-clear	Annular-clear
33	0.10	Annular	Annular-stratified	Annular-stratified	Annular-clear
33	0.08	Annular	Annular-stratified	Annular-stratified	Annular-clear
33	0.07	Annular	Annular-stratified	Annular-stratified	Annular-clear
33	0.05	Annular	Annular-stratified	Annular-stratified	Annular-clear
33	0.04	Annular	Annular-stratified	Stratified	Annular-clear
28	0.10	Annular	Annular-stratified	Stratified	Annular-clear
28	0.08	Annular	Annular-stratified	Stratified	Annular-stratified
28	0.07	Annular	Stratified	Stratified	Annular-stratified
28	0.05	Annular	Stratified	Stratified	Stratified
28	0.04	Annular	Stratified	Stratified	Stratified

24	0.10	Annular	Annular-strat-ified	Stratified	Annular-pseudo slug
24	0.08	Annular	Annular-strat-ified	Stratified	Annular-stratified
24	0.07	Annular	Stratified	Stratified	Stratified
24	0.05	Annular	Stratified	Stratified	Stratified
24	0.04	Annular	Stratified	Stratified	Stratified
19	0.10	Annular	Annular-strat-ified	Stratified	Annular-pseudo slug
19	0.08	Annular	Annular-strat-ified	Stratified	Annular-mist
19	0.07	Annular	Stratified	Stratified	Stratified
19	0.05	Annular	Stratified	Stratified	Smooth-stratified
19	0.04	Annular	Stratified	Stratified	Smooth-stratified

Effect of Molecular Weight

As in single-phase flow, the effectiveness of DRPs in multiphase flow is enhanced by higher molecular weight of polymer. Although very little attention has been paid to study the effect of this parameter on drag reduction, Al-Yaari et al. (2009) has carried out a study by injecting three polyethylene oxide polymer solutions, with identical chemical structures and concentrations but with different molecular weights of 3×10^5, 4×10^6 and 8×10^6 into water continuous dispersed flow regime. The results showed that drag reduction increased with increase in molecular weights.

Effect of Concentration

Generally, drag reduction increases with increase in the concentration before levelling off (Al-Sarkhi and Hanratty, 2001a, Al-Sarkhi and Hanratty, 2001b and Mowla and Naderi, 2006). Al-Sarkhi and Hanratty (2001a) have shown that the concentration of the polymer master solution largely influences the optimum drag reduction.

Mowla and Naderi (2006) showed that percentage drag reduction was affected by DRP concentration in the air–crude oil slug flow. There was a maximum (called critical concentration) above, which no more reduction could be obtained. This optimum concentration (18 ppm) was independent of the type of the pipe material or its diameter. This however contrasted the earlier result of air–water annular flow, which indicated that drag reduction depend on pipe ID (Al-Sarkhi and Hanratty, 2001b). Therefore, it can be said that drag reduction of DRPs also depends on the nature of the fluids in gas–liquid flow as well as the flow pattern.

Al-Yaari et al. (2009) injected 2, 5 and 50 ppm of Magnafloc 1101 into water continuous dispersed flow regime to study the effect of the concentration. They found that the concentration effect became very pronounced when the water volume fraction was increased beyond the phase inversion water fraction of 0.34. At the phase inversion point, the increase of the concentration from 2 to 5 ppm increased the pressure gradient reduction from 45 to 55% while with 50 ppm, the phase inversion point was shifted to a lower water fraction of 0.28. They attributed this shifting to be a result of a different flow pattern formed at this particular concentration. The flow pattern boundaries have also been found to be affected by the concentration of DRPs. This was investigated by Al-Wahaibi et al. (2007) who used 20 and 50 ppm Magnafloc 1011 in oil–water horizontal flow. The increase in the concentration has clearly extended the stratified region to a higher superficial water velocity (Fig. 13).

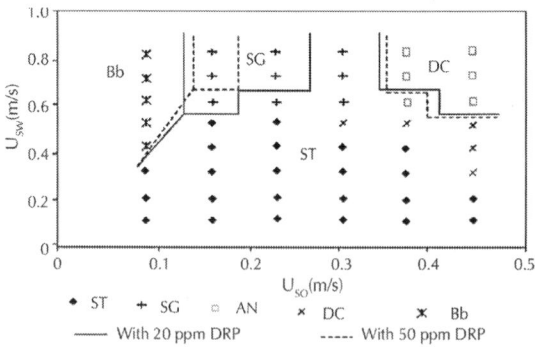

Figure 13: Comparison of flow pattern boundaries of oil–water flow without DRP against its flow pattern boundaries with 20 and 50 ppm DRP in

a 14 mm ID acrylic horizontal pipe. Flow patterns: ST (stratified smooth or wavy), DC (dual continuous), SG (slug), Bb (bubble) and AN (annular) (Al-Wahaibi et al., 2007).

Effect of Salts

Generally, an increase in salt concentration has a negative effect on the performance of water-soluble DRPs (Al-Yaari et al., 2009). This is attributed to the change in polymer conformation with the change in salt concentration due to electrolytic nature of the saline solution. Normally crude oil contain about 5% salt and this salt is capable of causing electrostatic interaction between different parts of a particular ionic polymer chain thereby leading to coiling of the polymer molecules. Such a situation hinders aggregate formation and expansion of the polymers, a phenomenon that is responsible for drag reduction (Karami and Mowla, 2012).

MEASUREMENT TECHNIQUES

Single Phase Flow

The measurement techniques involving DRPs in single-phase flow consist of pressure drop and velocity (or turbulence) measurements.

Pressure Drop Measurements

Drag reduction is usually obtained from the measurement of pressure drops of the fluid with and without DRPs. Various pressure transducers have been used to measure the pressure drop. For instance, wall-pressure fluctuations in turbulent channel flow were measured using four PCB A-1 16 piezoelectric transducers where each transducer was mounted in a cavity in order to minimize the effect of spatial averaging at high frequency by reducing the size of the transducer sensitive area to the pinhole cavity dimension

(Jourdan et al., 1998). Warholic et al. (1999) employed Validyne reluctance pressure transducer (Model DP103) to measure the pressure gradient while investigating the influence of DRP on turbulence. The transducer is located over a distance of 152 cm on the bottom wall of the channel. This type of transducer has also been used by Liberatore et al. (2004) for another flat wall with a separation distance of 86 cm to measure the pressure drop and the wall shear stresses for different amounts of drag reduction. The pressure drops in pipelines are also measured using differential pressure transducers where flow is already fully developed (Gasljevic et al., 2001). Ptasinski et al. (2001) used the membrane type with 88 mm water pressure full scale. Allowing the test section to be as long as 8 m to minimize the relative measuring errors, the measurement starts at 16 m (400 pipe diameters) after the entrance of the pipe to ensure fully developed flow by having constant pressure gradient in the entire segment. Similarly, Vanapalli et al. (2005), Japper-Jaafar et al., 2009 and Japper-Jaafar et al., 2010, Kim et al. (2009) and Gallego and Shah (2009) have all used differential pressure transducers from different manufacturers for polymer solutions in pipes.

Velocity Measurements

Apart from the fluid bulk velocity which is usually obtained from the measurement of flow rate using either automated or manual flow metres, the velocity profile and turbulent fluctuations are generally measured by either Laser Doppler Velocimetry (LDV) or Particle Image Velocimetry (PIV) technique. It has been discovered that the initial use of hot-film probe measurement are not reliable because non-Newtonian polymer solutions and water did not produce the same amount of heat and the accompanying shear stress. Therefore, the use of these Velocimetry techniques is necessary to determine accurate wall shear stresses and resulting drag reduction (White et al., 2004). Some of the authors that have used LDV in polymer solutions include Jourdan et al. (1998), Warholic et al. (1999), Ptasinski et al. (2001), and Japper-Jaafar et al., 2009 and Japper-Jaafar et al., 2010 while Warholic et al. (2001), Liberatore et al. (2004) and Motozawa et al. (2012) used PIV. However, the curved surface of the pipe can cause optical refraction and a small part can be replaced with a special test section to minimize this complication.

Multiphase Flow

Pressure Drop Measurements

The pressure drop measurement techniques employed for single phase flows are basically the same used for multiphase flows. For pressure drop determination of multiphase flows, some of the instruments which have been in use include U-tube manometer sometimes inverted or filled with Mirian blue fluid and various differential pressure transducers such as Validyne variable resistance, pressure sensors, and capacitance differential pressure transducer are used at different locations in the pipelines downstream from the injection points of the polymers (Kang and Jepson, 2000, Mowla and Naderi, 2006 and Wilkens and Thomas, 2007).

Flow Patterns Observation

In multiphase flows, the flow patterns are observed by visual observation, photographic or video techniques and probes. The use of visual observation methods in multiphase flow experiments to observe flow pattern variations is achieved by observing the flow through a transparent plexiglass or acrylic pipes, and by inserting a window or section on non-transparent pipelines. The extension of visual observation is the use of photographic or video methods. For very rapid phenomena, high-speed photography or video is used (Al-Sarkhi and Hanratty, 2001a, Al-Sarkhi et al., 2006, Al-Wahaibi et al., 2007, Al-Yaari et al., 2009 and Al-Sarkhi, 2012).

In a multiphase study without DRP, Angeli and Hewitt (2000) used a video recorder coupled with an endoscope attached to a camera to study dispersed phase sizes at different locations inside and along the pipe length. Because of the deficiency of the photographic or video method to accurately delineate flow pattern, as they only record the flow from outside the pipe, close to the pipe wall, the use of probes becomes necessary. Study of liquid–liquid flows without DRPs has particularly mentioned such measurement techniques as high frequency impedance probe, conductivity needle probe, dual impedance probe and hot-film anemometer (Xu, 2007). High frequency impedance probe is used to discriminate between the phases based on a variety of properties such as thermal conductivity, refractive index, electrical resistance or electrolytic current. Electrical impedance probes are very suitable for oil–water flows because of large difference in electrical properties between oil and water. In particular, alternating current high frequency impedance probes are most suitable for oil–water flow measurement because of the large dielectric constant between the two liquids. Conductivity needle probe is particularly used to measure what is called the Phase Inversion Point (PIP) in the dispersed flow regime which is not readily identified since one phase (water) is conductive and the other (oil) non-conductive. The dual impedance probe is an impedance probe with two sensors used specifically to measure drop velocity and drop size distribution in multiphase flows. According to Angeli and Hewitt (2000), conductivity probe has also been used to identify the continuous phase in the dispersed region while impedance probe was used to obtain phase distribution in inclined oil–water pipeline flow.

Liquid Hold up Measurements

The liquid hold-up are determined by using conductivity probes of different configurations, quick closing valves or by analysing the pictures obtained from high speed imaging system (Al-Sarkhi and Hanratty, 2001a, Soleimani et al., 2002, Baik and Hanratty, 2003, Al-Sarkhi and Soleimani, 2004, Al-Wahaibi et al., 2007, Al-Yaari et al., 2009, Al-Sarkhi, 2012 and Edomwonyi-Otu et al., 2013). The probes are usually calibrated by measuring the conductance of the fluid in the pipe at different levels and a plot of the ratio of the hold-up to pipe ID against the conductance will give a linear relationship which enables time-average liquid height to be determined from time-average conductance. In addition, some hot-film anemometers are capable of separating the total signal into the parts corresponding to bubbles and continuous phase, from which the local volume fraction can be calculated.

INJECTION METHODS OR EFFECT OF MIXING

There are basically two ways of adding polymers to a flowing fluid: homogeneous and heterogeneous methods. While in homogeneous method the polymer powder is dissolved directly into the fluid and allowed to mix for several hours before the fluid is pumped through a pipe, the heterogeneous method involves injecting polymer solution into the fluid flow after the pump (Smith and Tiederman, 1991). The injection of the polymer solution using heterogeneous approach can also be achieved by two ways. The first is called diffusing injection where the polymer solution is injected into the centre or at the wall of a pipe at a concentration which disperses completely by turbulent diffusion so that at some distance downstream of the injection site, the solution is completely homogeneous and well mixed. Polymer thread injection is the second way where a high concentration of polymer solution is injected into the centre or at the wall of a pipe flow such that a single, coherent, unbroken polymer thread which begins at the injection site is produced and continues downstream for several hundred pipe diameters (see Fig. 14a–c).

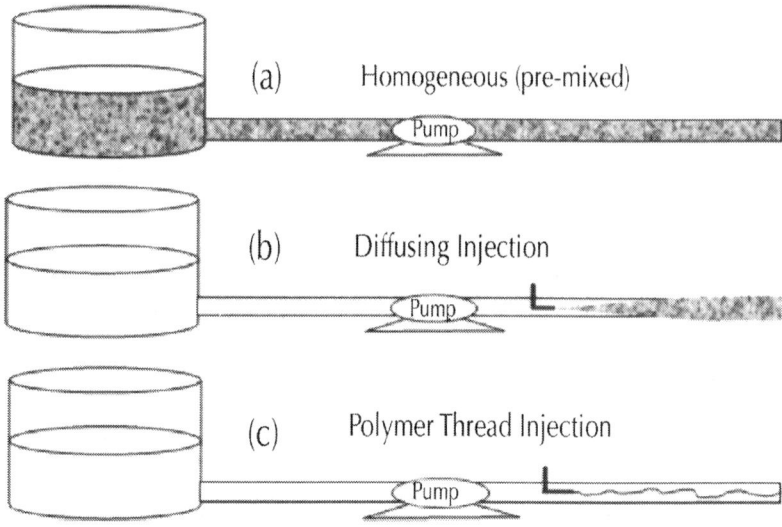

Figure 14: Schematic representation of three different approaches for using polymer drag reduction (Smith and Tiederman, 1991).

The studies of Smith and Tiederman (1991), Vlachogiannis et al. (2003) and Baik et al. (2005) have shown that the heterogeneous polymer injection significantly enhances the drag reduction effectiveness (can even be doubled for a given average polymer concentration) when compared to the homogeneous polymer addition. This is the reason why the heterogeneous method, particularly the diffusing injection method, has been employed for most polymer drag reduction investigations. Wyatt et al. (2011) took advantage of this to suggest pseudo-heterogeneous method to study the effect of polymer entanglements in drag reduction. In their proposal, an entangled stock solution is prepared in a holding tank and allowed to completely hydrate for a sufficient time after which it is diluted to a desired polymer concentration. The diluted solution is then immediately injected into a turbulent pipe flow without further resting time. This new method of injecting polymer solution provides an opportunity to study the effects of residual entanglements and initial polymer structure on drag reduction effectiveness.

Single Phase Flow

In most experimental investigations, the fluid and the polymer additives were usually premixed in storage tanks using different mixer and then the mixtures were pumped into the flow loop (Warholic et al., 1999, Gaard and Isaksen, 2003 and Kim et al., 2009). This method of introducing the polymers can result to mechanical degradation thereby reducing the effectiveness of the polymers. Better injection method involved the use of three-hole nozzles at low pressure that is oriented in the vertical and ±inclination angles from the vertical. This enhances the formation of symmetrical polymer film around the whole pipe circumference which reduces the skin friction with greater extent and promotes modified turbulent flow with lower friction factor or pressure drop as compared to premixing method or even one-hole nozzle injection method (Fig. 15). For instance, Jha et al., 2012a and Jha et al., 2012b, conducted an investigation on the drag reduction of polyacrylamide solution using one-hole and three-hole nozzles at different polymer solutions in a pipeline of 25.3 mm ID and 10.2 m long. It was discovered that the three holes polymer injection nozzle showed higher efficiency of drag reduction of 44.9% at 10 ppm than single holes polymer injection nozzle system (41.2%) at the same polymer concentration. In addition, drag reduction was found to decrease with the increase in the polymer solution for both injection systems.

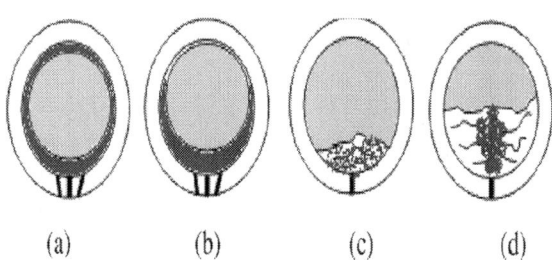

(a)   (b)   (c)   (d)

Figure 15: Polymer film distribution in pipeline (a) 3-hole nozzle at low pressure, (b) 3-hole nozzle at high pressure, (c) single-hole nozzle at low pressure, and (d) single-hole nozzle at high pressure (Jha et al., 2012a).

Multiphase Flow

Generally, the injection of polymer solution into a multiphase flow can be achieved using a low shear pump as in the studies of Al-Wahaibi et al. (2007) and Yusuf et al. (2012), or using a pressurized tank to avoid polymer degradation (Al-Yaari et al., 2009). However, the effectiveness of any polymer is dependent on the injection method used to introduce it into the flow. For example, several studies have shown that master solution of DRP need to be injected in a way that the polymer solution is distributed across the whole circumference of the pipeline. To buttress this point, Al-Sarkhi et al. (2006) introduced the polymer master solution through a hole of 2 mm diameter which was located at the bottom of the pipes at 1.0 m downstream of the mixing tee of air and water where the annular flow was fully developed. They found that the polymer was rapidly mixed with the liquid flowing along the pipe wall. Three-hole injection method has also been applied instead of this one-hole injection system to introduce the polymer master solutions into the flow in some experimental investigations and found to enhance better distribution of the polymer solutions (Al-Sarkhi and Hanratty, 2001a, Soleimani et al., 2002 and Al-Sarkhi and Soleimani, 2004). The three holes were oriented in such a way that one of the holes is vertical and the other two were ±15° from vertical. The vertical jet feeds the liquid at the very bottom of the pipe while the inclined jets feed on both sides of the pipe.

The injection point of the polymer master solution into the flow is another important factor that has been investigated. Injection of DRP may be carried out downstream in which case we have the Downstream Injection Point (D.I.P) or at the entrance in which case we have the Entrance Injection Point (E.I.P). Al-Sarkhi and Hanratty (2001a) and Al-Sarkhi (2012) investigated the effect of injection point on drag reduction. They both found that drag reduction is larger when DRP is injected downstream. In addition, the onset of drag reduction in the case of downstream injection starts at lower polymer concentration while the effectiveness of DRP injected downstream is not affected by gas velocity unlike the entrance injection where the effectiveness decreases with increasing gas velocity. They explained that this might be due to structural arrangement of the polymer aggregates when they pass through the small holes through which they are introduced in the case of downstream injection method. For the entrance injection, there might be impingement of air jet on the mixing tee which will result into the breaking up of DRP aggregates or a change in the structural arrangement of the DRP. Hence, higher concentration DRP are required for entrance injection to have the same effectiveness as the downstream injection.

THEORETICAL STUDIES OF DRPS IN PIPELINES

Experimental studies of DRPs in turbulent flows have been given much attention yet a comprehensive theory of polymer drag reduction which can predict the magnitude of drag reduction in a given turbulent flow based on the polymer properties and can explain some important factors that influence drag reduction is still not fully known. While most popular theory attributes drag reduction to the enhanced elongational viscosity of the polymer solution others ascribe it to elastic theory, viscoelasticity, or stress anisotropy of the polymer solution.

Single Phase Flow

Lumley (1969) was the first to observe that there is an overlap in time-scales of polymer and turbulence dynamics but their length-scales do not overlap. The maximum length of polymer molecules according to him can range from 1 ppm to 1 nm whereas the turbulence length-scales in high Reynolds number boundary layer can be as small as 0.1 μm but not much smaller. On the other hand, comparing the time-scales of polymer motions (which are in the range of 10^{-12} and 10^{-2} s and even larger) with turbulence motions (which can be as small as 10^{-5} s and as large as minutes and above) showed that there is an overlap in the range 10^{-5} to may be up to a second. This overlap in the time-scales necessitates the importance of Weissenberg (or Deborah) dimensionless number (Eq. (9)) which is defined as the ratio of the polymer relaxation time (τ) to a time scale of smallest turbulent eddies (tf). This dimensionless number, however can be greater or <1, but drag reduction is only noticed when it is significantly >1.

$$De = \frac{\tau}{t_f}$$

(9)

Further study carried out by Lumley (1973) postulated that the increase of kinematic viscosity as a result of polymer additives causes the Kolmogorov length-scale , which in a turbulent boundary layer depends on the distance to the wall, to increase. Therefore the buffer layer is in turn increased and by implication the drag reduction. He postulated further that the elongation of polymer molecules is due to strain flow i.e. the elongation of flow component, and that vorticity reduces polymer elongation because it shortens the relative time that the elongated polymer is aligned with the principal axis of the strain field. Based on this mechanism, he qualitatively explained the maximum drag reduction phenomenon of Virk as a feedback saturation effect of turbulence causing polymer molecules to stretch, which in turn suppress turbulence.

Landahl (1977) developed a two-scale mechanistic model of turbulence based on the classical hydrodynamic stability concept. He argued that shearing of large scale eddies by mean flow can change the local velocity profile, leading to small scale instability, which in turn can further develop into a bursting event. Ryskin (1987) produced a modified version of the Lumley›s model. In his "yo-yo" model of describing polymer in an extensional flow, he scaled relative drag reduction with effective viscosity increase and his prediction were in general agreement with Virk. In contrast with Lumley›s model which was based on viscosity effect, De Gennes (1990) proposed a model based on elastic properties of the solute polymer in which turbulence is suppressed by elastic absorption of moderately stretched polymer molecules and polymer elongation is thought to be proportional to a power law of length-scale ratio. He explained that the shear waves which are caused by the elasticity of the polymers prevent production of turbulent velocity fluctuations at the small scales.

The interactions between turbulence and polymer were considered by Bhattacharjee and Thirumalai (1991) using a simplified model for time response of the polymer molecules. They suggested that flexible polymers lead to an enhancement of molecular viscosity at small scales and hence the turbulence at such small scales is damped unlike at intermediate length scales where the effective viscosity is decreased.Sreenivasan and White (2000) used the de Gennes descriptive model to calculate on-set shear stress by defining the on-set point as a scenario in which de Gennes threshold length scale is reached at the buffer layer between the viscous wall-layer and the overlap layer. They further extended de Gennes model to determine the condition for which MDR is reached by assuming that one of the conditions that define the point of MDR is when the polymer concentration reaches dilute solution limits i.e. when the polymer molecules start to overlap.

Ptasinski et al. (2001) have shown that polymer chains are significantly suppressed despite the fact that the polymer chains themselves contributes to the extra stress tensor in the momentum equation. These studies revealed that disentangling of polymer chains is actually an essential requirement in the ability of a polymer to reduce viscous drag. In another fascinating work by Ptasinski et al. (2003), they determined the turbulent kinetic energy budgets and concluded that a major part of the energy production by the mean flow is transferred directly into the elastic energy of the polymer chains. The energy gained by the polymer macromolecules is partly contributed by the turbulent velocity fluctuations and their analysis revealed that polymers dissipate these energies through a relaxation process with the highest dissipation taking place in the buffer layer.

Pinho (2003) modified the generalized Newtonian fluid (GNF) model to account for strain-thickening of the extensional viscosity derived equations for mass, momentum, Reynolds stresses, turbulent kinetic energy and its rate of dissipation. Using these equations as governing equations, he proposed a closed model for the time-average viscosity that takes into account its non-linearity and dependence on the second and third invariants of the fluctuating rate of deformation tensor. Cruz et al. (2004) proposed a model for the new stress term in the momentum equation of a generalized Newtonian fluid that is used to mimic viscoelastic effects of the fluid exhibiting drag reduction in turbulent pipe flow. The new stress quantifies the cross-correlation between the fluctuating viscosity and the fluctuating rate of strain that had been neglected in the $k-\epsilon$ low Reynolds number model originally developed by Pinho (2003). They stated that with the inclusion of the new stress, the prediction of turbulent kinetic energy in drag reducing pipe flow is significantly improved.

L'vov et al. (2004) developed a turbulent boundary layer model based on the momentum balance (between viscous shearing stress, Reynolds stress, and pressure) in combination with an approximated turbulent energy equation (a balance between turbulent energy production, and its estimated dissipation). They obtained a logarithmic velocity profile by defining a constant ratio of Reynolds stress to turbulent kinetic energy. In addition, they obtained the maximum drag reduction asymptote by a closure relationship where it is assumed that the Reynolds stress is proportional to the kinetic energy with a constant of proportionality that tends to zero as the polymer concentration increases, and a kinematic viscosity that is a function of distance y from the wall as a result of polymer addition.

Resende et al. (2006) developed an anisotropic low Reynolds number $k-\epsilon$ turbulence model and compared its performance with experimental data for fully developed turbulent pipe flow of four different polymer solutions. According to them, the prediction of friction-factor, mean velocity and turbulent kinetic energy show only slight improvement over those of a previous isotropic model. The new turbulence model is capable of predicting the enhanced anisotropy of the Reynolds normal stresses that accompanies polymer drag reduction in turbulent flow. Mehrabadi and Sadeghy (2008) later discovering that the time-averaged turbulence formulation derived by Pinho (2003) and Cruz et al. (2004) requires a low-Reynolds number k–☐turbulence model to function used "Launder-Sharma" k–☐ model (Launder and Sharma, 1974) to see the possibility of predicting the huge drag reduction which has been observed for several polymer solutions. It was found that the performance of this model on f–Re curve is better than the one used by Cruz et al. (2004), possibly because of the more suitable adjusting parameter obtained.

Sher and Hetsroni (2008) modelled drag reduction in walled turbulent flows of polymer solutions using a mechanistic model. In this mechanistic model, which is a force balance on a polymer molecule in a turbulent flow field, elastic and centrifugal forces were taken as dominant forces on a polymer fibre in the turbulent flow field. They argued based on the model that viscous elongation of the polymer is negligibly small due to high eddy frequency and a relatively slow typical polymer response.

Iaccarino et al. (2010) proposed a novel eddy viscosity model for predicting friction drag reduction induced by polymers in turbulent wall-bounded flows. This model, according to them is based on the elliptic relaxation model modified to account for the modified Reynolds-stress equilibrium established by the presence of elastic polymer chains in the fluid. They claimed that the model reproduces the level of drag reduction observed over a wide range of rheological parameters and that both the mean velocity and turbulent fluctuations are predicted with good accuracy. Resende et al. (2011) proposed a new low Reynolds number k–☐ turbulence model for flows of viscoelastic fluids described by the finitely extensible nonlinear elastic rheological constitutive equation with Peterlin approximation (FENE-P model). The model which represents an improvement over the Pinho et al. (2008) model also compares favourably against direct numerical simulations in the low and intermediate drag reduction regimes of up to 50%.

Some empirical correlations are also available in the literatures. For instance Sood and Rhodes (1998)developed a pipeline scale-up model in drag reducing turbulent by equating dampening of turbulent velocity fluctuations by drag reducing additives to a reduction in the Prandtl mixing length. With the developed correlation (Eq. (10)), the two constants, K and B are determined from average bulk velocity obtained in laboratory small scale pipe diameter and thereafter used to predict average bulk velocity of larger pipe diameter.

$$\bar{u}^+ = \frac{1}{K} \ln R^+ + B$$

(10)

With

$$B = \frac{A - 15}{K}; \quad R^+ = \frac{Ru_\tau}{\vartheta}; \quad \bar{u}^+ = \frac{\bar{u}}{t_\tau}$$

where K, Prandtl's mixing length constant; R, pipe radius; u , wall shear stress; ū, average bulk velocity; , kinematic viscosity; and A, constant.

This model was successful in predicting flow in a Library cooling system 154-mm ID pipe from experimental data obtained in a 7-mm ID tube (water–surfactant system) and in the Trans Alaska 1194-mm ID pipe from experimental data obtained in 26.6-mm ID pipe (oil–polymer system). This showed that the model is not restricted in application to a particular system and it is very easy to use because it does not require measurement of the troublesome characteristic time or the velocity shift correlation – a small laboratory scale pipe providing flow rate, pressure drop data and viscosity is enough to predict flow in bigger pipes. Gallego and Shah (2009) made the first attempt to obtain a generalized drag reduction correlation in terms of Deborah number for flow of drag reducing polymer solutions in coiled tubing (Eq. (11)) in additon to straight tubing (Eq. (12)). Experimental data of polymer solutions at several concentrations and temperatures flowing through straight tubing and coiled tubing sections of different diameters, curvature ratios, and roughness were used in developing the model. It was found that the predictions from the friction pressure correlation for flow in both straight and coiled tubing are in reasonably good agreement with the experimental data.

$$N_{De} = \frac{1.1861(f_sN_{Res})^{0.3221}(8v\gamma/d)}{[1 + (8v\gamma/d)^2]^{0.7034}} \left(\frac{\rho_p\mu_s}{\rho_s\mu_o}\right)^{0.1918}$$

$$(22,000 \leq N_{Res} \leq 300,000)$$

(11)

$$N_{De} = \frac{1.6675 \times 10^{-3}(f_sN_{Res})^{1.4084}(8v\gamma/d)}{[1 + 1.0974 \times 10^{-3}(f_sN_{Res}(8v\gamma/d))^{1.42305}]^{0.75110}} \left(\frac{\rho_p\mu_s}{\rho_s\mu_o}\right)$$

$$(22,000 \leq N_{Res} \leq 430,000)$$

(12)

where N_{De}, Deborah number; f_s, fanning friction factor of the solvent; N_{Res}, solvent Reynolds number; γ, relaxation time (s); μs, solvent shear rate viscosity (cP); μo, zero shear rate viscosity; ρp, polymer solution density (lb/gal); ρ_s, solvent density (lb/gal); v, average fluid viscosity (ft/s); and d, pipe inner diameter (in.).

The Deborah number and the fanning friction factor are related to drag reduction (DR) as presented in Eqs.(13) and (14).

$$N_{De} = \left(\frac{f_s}{f_p}\right)^2 - 1$$

(13)

$$DR = 1 - \frac{f_p}{f_s}$$

(14)

where f_p, fanning friction factor of the polymer solution.

Numerical simulations have also been combined with models for the polymer molecules over the years for the purpose of studying drag reduction. Due to the difficulties associated with making laboratory measurements of the polymer configuration in turbulent flow, it has not been possible to test the theories highlighted above by direct experimentation. However, recent breakthrough particularly in direct numerical simulations of turbulence, combined with continuing developments in computer power, have made it possible to begin to address some of these problems through numerical simulation.

Bird et al. (1987) used the Finitely Extensible Non-Linear Elastic (FENE) model which is based on an elastic dumbbell that depicts the polymer molecule as two beads linked by a spring. The spring forces which are modelled by a nonlinear equation tend to return the polymer back to its coiled form. The non-linearity of the spring ensures that the dumbbell cannot exceed more than a given maximum extension. Although FENE-P model which contains an approximation Peterlin derived parameter is a modified form of FENE model by Peterlin (1966), the majority of these studies which involves the dynamics of the polymer in turbulent flow has been represented by a FENE-P dumbbell or chain model (Massah et al., 1993, den Toonder et al., 1995, Sureshkumar et al., 1997, Dimitropoulos et al., 1998, Baron and Sibilla, 1998,Massah and Hanratty, 1997, Kumar and Homsy, 1999, Min et al., 2001, Ilg et al., 2002 and Vaithianathan and Collins, 2003). This is because the FENE-P model is attractive in the sense that it is based on a closed constitutive equation derived from the original FENE model using a self-consistent pre-averaging approximation and the closed constitutive equation significantly reduces the computational cost of the FENE-P model compared to its FENE counterpart, which can only be implemented using expensively stochastic simulations. Again, while the FENE-P model has been able to provide a good approximation to the FENE model in steady elongational flow and in steady shear flow at low shear rates its predictions in time-dependent elongational and shear flows deviate both from experiments and the predictions of the original FENE model (Zhou and Akhavan, 2003). Recently, CFD simulations were performed on fully stratified gas shear-thinning liquid and gas shear-thinning liquid slug flow regimes in order to compare the drag reduction ratio and pressure gradients. There was a great agreement of pressure gradients with experimental data for wide gas flow rates (Jia et al., 2011).

Multiphase Flow

Only the theoretical studies of single phase polymer solutions have been mostly focused with little emphasis in the case of multiphase flows. However, very few reported studies are found with regard to empirical correlation of multiphase flow with DRPs. For instance, Daas et al. (2000) developed physical models to predict the frictional and accelerational components of the total pressure gradient in a horizontal gas–liquid slug flow (Eqs. (15) and (16) respectively). They found that the predicted values agreed well with the measured values except at low gas velocities and at conditions near to the stratified/slug transition.

$$\Delta P_{f.body} = \frac{2f_{slug}V_s^2[\rho_L R_s + \rho_G \alpha_s](I_s - I_{MZ})}{D}$$

(15)

$$\Delta P_a = \rho_L A R_f(V_s - V_f)$$

(16)

where f_{slug}, slug friction factor; V_s, average no-slip velocity of the fluid in the slug body; V_f, liquid film velocity; R_s, liquid holdup in slug; R_f, liquid holdup in the stratified film; Is, length of slug body; IMZ, length of mixing zone; A, area; ρ_L and ρ_G, density of liquid and gas respectively.

Fernandes et al. (2004) equally developed a model for the variation of drag reduction with superficial liquid velocity, superficial gas velocity and pipe diameter in horizontal annular two-phase flow. The model which accounts for the drag reduction as a reduction of the height of the short-wavelength waves on the liquid film and a reduction of the entrainment rate of droplets from the liquid film into the gas core (Eq. (17)) produces a reasonable quantitative agreement between the model predictions and the present experimental data.

$$\frac{dP}{dx} = -\frac{4}{D}\tau_{fr} - \frac{4}{D}E_r(U_{SG} - U_i)$$

(17)

where τ_{fr}, interfacial friction shear stress; E_r, entrainment rate; USG, gas superficial velocity; and U_i, velocity of interface.

The first term on the right hand side of Eq. (17) represents the pressure gradient due to interfacial friction while the second term represents that due to entrainment. Both of the terms are quantitatively determined from earlier correlations of Wallis (1969) and Schadel et al. (1990) respectively.

Recently, Al-Sarkhi et al. (2011) developed two correlation models to predict the effect of DRP on gas–liquid (Eq. (18)) and liquid–liquid (Eq. (19)) flows at maximum drag reduction.

$$f_{M\_DRP} = 3.36 \times 10^{-7} \frac{D_0}{D} \left(Re_M \left(\frac{U_{SG}}{U_{SL}} \right)^{0.5} \right)^{0.595}$$

(18)

$$f_{M\_DRP} = 0.614 \left(Re_M \left(\frac{U_{SW}}{U_{SO}} \right)^{0.5} \right)^{-0.5}$$

(19)

where f_{M_DRP}, fanning friction factor with DRP; DO, reference pipe diameter; D, experimental pipe diameter; U, superficial phase velocity.

The models which involved the friction factor of the two multiphase flows with DRP at maximum drag reduction as a function of mixture Reynolds number and the ratio of the superficial velocities were found to be applicable for wide range of pipe diameters based on the results of the published data for air–liquid annular flow and oil–water flow of any flow pattern with DRP in pipes.

SUGGESTED MECHANISMS BEHIND DRAG REDUCTION WITH DRPS

Single Phase Flow

Drag reduction and flow pattern modification by polymer molecules is a very complex process. To explain the mechanisms behind this phenomenon, researchers have propounded several theories. All of these are semi-empirical or highly speculative, and all have been subject of criticism. Oldroyd (1949) proposed the first explanation for drag reduction by polymeric molecules. He argued that an external constraint is imposed by the wall on the ways in which long-chain molecules can rotate near the wall; therefore, an abnormally mobile laminar sub layer, of thickness comparable with molecular dimensions can exist at the wall, which causes an "effective slip". Elperin et al. (1967) suggested that an adsorbed layer of polymer molecules could exist at the pipe wall during flow and this could lower viscosity, create a slip, damp turbulence and prevent the formation of vortices at the wall. Some theories were based on the interaction between the polymer and the turbulence using the existence of a drag reduction onset Reynolds number as an indicator of this interaction. Virk et al. (1967) postulated that onset occurs if the ratio between the turbulence and polymer length scales reaches a critical value. After onset, some interaction between the polymer molecules and turbulent flow results to give drag reduction.

The mechanism based on the extension of the polymers as proposed by Lumley, 1969, Lumley, 1973 and Lumley, 1977 represents one of the two major theoretical concepts being put forward to explain the phenomenon of drag reduction by polymers. He postulated that stretching of randomly coiled polymers – primarily in regions with strong deformations such as the buffer layer – increases the effective (extensional) viscosity, resulting in damping of small eddies, thickening of the viscous sub-layer and consequently bringing about drag reduction (see Fig. 16). Lumley also mentions that the influence of the polymers on the turbulence only becomes important when the time scale of the polymers (e.g. the relaxation time) becomes larger than the time scale of the flow, which is known as the onset of drag reduction.

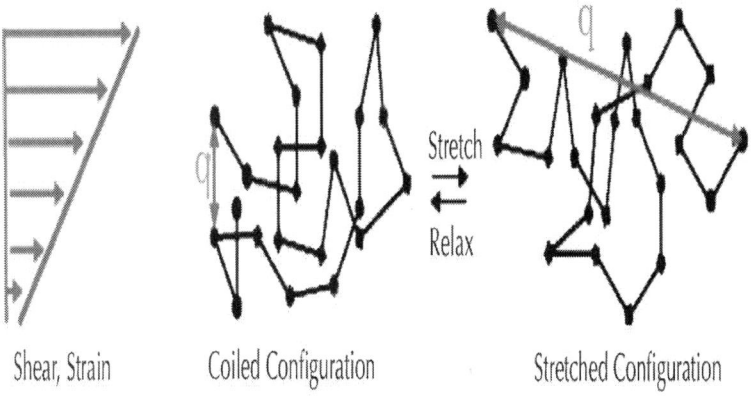

Shear, Strain   Coiled Configuration   Stretched Configuration

Figure 16: Schematic of polymer stretch and relaxation in shear flow (White and Mungal, 2008).

The experiments conducted by McComb and Rabie, 1979 and McComb and Rabie, 1982 where they found drag reduction also occurring when the polymers have been injected in the centre of the pipe, lend credence to the second major theory proposed by De Gennes (1990) that drag reduction is caused by the elastic rather than the viscous properties of polymers. Tulin (1984) proposed that DR effect seems to depend on the stretching of individual molecules by high strain rates in flow. He explained further that at high strain rates, polymer chain tends to elongate along the principal strain rate axis resulting in large extensions. At the same time, a form of strain-rate hardening occurs in which elongation viscosity becomes very high. As the elongation velocity increases, the large scale bursts and the sweeps in the wall layer flows are inhibited, thus reducing friction.

Massah et al. (1993) computed the behaviour of a single polymer molecule in a turbulent channel flow. They concluded that it is plausible to associate drag reduction with pre-shearing of the polymers in the viscous sublayer, as well as with large temporal changes of polymer configurations in the buffer layer.Handler et al. (1993) obtained drag reduction in a simulated channel flow by randomizing phases of some of the Fourier modes. Much of the results they obtained agreed with experimental results on polymeric drag reduction, however, some significant differences led them to conclude that drag reduction due to phase randomization may be due to the destruction of coherence of the turbulence producing structures near the wall. Using a finite difference based model to describe polymer drag reduced flow in a pipe, Baron and Sibilla (1997) proposed that polymers reduce drag by becoming elongated and thus inhibiting turbulence regeneration by opposing pressure redistribution from stream-wise to cross-flow velocity fluctuations.L'vov et al. (2005) suggested that drag reduction may be due to the generation of high quality hydro-elastic waves as a result of the conversion of turbulent energy into polymer potential energy and vice versa.

Dubief et al., 2004 and Dubief et al., 2005 used numerical experiments to investigate certain features of polymer–turbulence interaction. They found that polymers reduce drag by damping near wall vortices and sustain turbulence by injecting energy into the stream-wise velocity component in the very near wall region. In order to substantiate the works of Lumley, Absi et al. (2006) carried out study on the extensional viscosity of elastic liquids under strong flows and suggested that anisotropic limit of the extensional viscosity caused by the extended polymers should play a key role in the attenuation of flow instability and in the mechanism of drag reduction by polymer additives.

They found that the macromolecule elasticity becomes irrelevant in this nonlinear flow regime and the flow depends mainly on the anisotropy produced by the alignment of the highly extended macromolecules with the flow direction. This is the reason why few ppms of polymers of high molecular weight probably have the ability to produce drastic reduction in the pressure drop of turbulent flows.

Multiphase Flow

The lack of knowledge on the understanding of the mechanism of drag reduction is most felt in multiphase flow because very little or no work has been done in this area. It is generally believed that in addition to the effects of DRPs on single-phase flows such as dampening turbulent bursts, wall roughness reduction and pipe wall wettability reduction, DRPs has numerous additional unique effects on multiphase pipe flow that lead to drag reduction. These include interfacial stress reduction, holdup change, flow pattern change, and reduction of effective density for vertical flows. However, some suggested explanations that could possibly shed light on the mechanism of drag reduction in multiphase flows by polymers can be found in some documented studies. Soleimani et al. (2002) in their study of drag reduction on pseudo-slug flow attributed the reduction to the destruction of the turbulence in the slugs and damping of waves at the interface by the polymer. They suggested that laminarization of the slug is associated with an increase in the velocity of the bubble behind a slug and this causes an increase in the shedding rate and, therefore, a destabilization of the slugs.

Fernandes et al. (2004), sharing the same view with Al-Sarkhi and Hanratty (2001a), assumed that the reduction of the height of short-wavelength waves on the annular liquid film which act as roughness elements and the reduction of the entrainment rates from the liquid film into the gas core are responsible for the drag reduction. In addition, drag reduction is majorly achieved from the suppression of the interaction between the gas and liquid phases in annular flow rather than the suppression of turbulence in the liquid film.

While most studies agreed that drag reduction by polymer is caused by the damping interfacial waves and by reducing the turbulence in the liquid (Baik and Hanratty, 2003), Al-Yaari et al. (2012) added that the main rheological property of the dilute polymer solution which is linked with drag reduction in single phase flow (i.e. elongational viscosities) also playing a significant part in the drag reduction of multiphase flows.

MAJOR PROBLEMS WITH THE USE OF DRPS

Degradation

There are two major problems that affect the effectiveness of DRPs when they are in continuous usage as drag reducers in turbulent flows for a certain period of time. These are dry out and mechanical (or shear) degradation. Dry out usually occurs in high gas–liquid ratio multiphase flow when the co-solvents for the dissolution of the polymers evaporate thereby leading to precipitation of the polymer and a loss of drag reduction benefit. Mechanical degradation on the other hand occurs under high shear conditions as a result of interactions with turbulence or being passed through a centrifugal pump. The long polymer chains are then broken permanently, thus leading to permanent reduction or even elimination of the drag reduction capabilities of the polymers. This mechanical degradation in turbulent flow is known to be dependent on various parameters, including polymer molecular weight (MW), molecular weight distribution (MWD), temperature, polymer–solvent interactions, polymer concentration, turbulence intensity, method of preparation and storage, and flow geometry.

Many interesting works have been carried out to investigate the degradation of DRPs. One of such is the work of Moussa and Tiu (1994) who investigated the effect of some of the above parameters and concluded that concentration and molecular weight delay the observed effects of degradation, thus increasing the critical Reynolds number at which degradation becomes pronounced. Choi et al. (2000)investigated the single exponential decay model through experimental data and showed that the model is not universally suitable for all DRPs except for some shear resistant polymers applicable in describing short time degradation behaviour. Kim et al. (2000) studied mechanical degradation of high molecular weight polystyrene under turbulent flow in a rotating disc apparatus using three different polymer solvents. As usual, degradation was observed with time while the extent of degradation was found to be influenced by the solubility parameters of the co-solvents. In addition, the theoretical model for molecular degradation in turbulent flow proposed by Brostow et al. (1990) was found to represent the experimental data quite well.

During polymer chain scission or breakage in turbulent flow, MWD was found to stabilize or reach a steady state after a long time of degradation. This finding was supported in studies using Poly -olefin in Varsol 80 (Nakken et al., 2001), Poly-isobutylene in kerosene (Lee et al., 2002), and PEO and PAM solutions (Vanapalli et al., 2005). They ascribed this to the fact that the mean molecular weight reaches an asymptotic value after a long time of the degradation process. The effect of temperature on degradation has been found to be very complex in the sense that there is a critical temperature below which degradation is delayed with increase in temperature and above which degradation increase with increase in temperature (Hadri et al., 2011). Investigations have revealed that higher concentration and molecular weight DRPs are less prone to degradation than their lower molecular weight counterpart as is evidenced from Shanshool et al. (2011) where they subjected polyisobutylene solutions of various concentrations and molecular weights to the same degradation before they were used for drag reduction. They discovered that for 10, 30 and 50 ppm solutions, drag reduction decreased from 7 to 0 after 60 min, 15 to 0 after 120 min and 21 to 9 after 180 min respectively. Also for molecular weights of 2.5×10^6, 4.1×10^6 and 5.9×10^6, drag reduction decreased from 9 to 0 after 60 min, 12 to 2 after 150 min and 21 to 10 after 180 min respectively.

Useful empirical models were recently developed to further address polymer degradation in turbulent flows. Hénaut et al. (2012) using their experimental data generated both from rheometer and small scale turbulent flow loop developed kinetic model with fitting parameters to analyzed the mechanical degradation due to high extensional rate resulting from bursts in relation with dissipated energy and drag reduction efficiency. On the other hand, Pereira and Soares (2012) proposed an alternative decay function that relates DRP as a function of the Reynolds number, concentration, molecular weight, and temperature using degradation result obtained in cylindrical double gap rheometer device.

However, different methods of applying DRPs in turbulent flows have been used to reduce the shear degradation. For instance injection of the polymers downstream of the pump and use of single pass flow loop (Rosehart et al., 1972), pumping of fluid from an open reservoir by a disc pump and other special pumps (Ptasinski et al., 2001) and pressure vessel instead of centrifugal pump to drive the flow (Zadrazil et al., 2012) have reduced shear degradation. It has also been found that the use of graft copolymers enhances shear stability (especially against biodegradation) (Deshmukh et al., 1991).

Toxicity

There is no documented work on the toxicity of DRPs in open literatures. However, it is a well-known fact that oil-soluble polymers are toxic because of their non-biodegradability while water-soluble polymers are non-toxic. Therefore it is advisable that before the discharge of used polymer, it must be made bio-degradable to avoid any toxic effect of the polymer on the environment.

APPLICATIONS OF DRPS

The major application of DRPs is the frictional drag reduction of turbulent flows which leads to savings in energy consumption. So it is possible to maintain the same flow rate using less power or to increase the flow rate or pipe length using the same amount of power. They equally serve as a modification of flow patterns especially in multiphase flows thereby enhancing separation of the phases. In addition, their applications can be found in numerous other fields involving fluid flow or transport. For instance, in crude oil pipeline flow, slurry or hydraulic capsule pipeline transportation, prevention of overflows of water in sewage systems after heavy rains (Dembek and Bewersdorff, 1981), reducing the self-noise of submarines and torpedoes (Truong, 2001), increasing water flow rate in fire-fighting equipment, waterborne shipping, water supply and irrigation systems, cooling and heating circulation system as well as in the improvement of blood flow for treating circulatory diseases like suppression of atherosclerosis (Mostardi et al., 1978 and Unthank et al., 1992), and lethality from haemorrhagic shock (Eichelberger, 1992, Kameneva et al., 2004 and Wang et al., 2011).

COMMERCIALLY AVAILABLE DRPS

Several DRPs have been mentioned as drag reducers in the literature. Basically, they can be classified based on the polarity of the solvents in which they are dissolved as shown in Table 4. Apart from the fact that some of them are available as homopolymers while others are produced as copolymers, many of the commercially available DRPs are represented by their trade names rather than their chemical names.

Table 4: Some commercially available DRPs

Water-soluble DRPs	Oil-soluble DRPs
Polyethylene oxide (PEO)	Poly isobutylene (PIB)
Polyacrylamide (PAM)	Polystyrene (PS)
Guar gum (GG)	Poly methyl methacrylate (PMMA)
Xanthan gum (XG)	Poly dimethylsiloxane (PDMS)
Carboxymethyl cellulose (CMC)	Poly cis-isoprene

Hydroxyethyl cellulose (HMC)	CDR*
PolyOx*	FLO*
SEPARAN* AP-30	Poly(1,2-butyleneoxide)
EP* 1000	Cis-poly butadiene
MAGNAFLOC 1011	Ethyl cellulose
MAGNAFLOC 1035	Copolymer of Epichlorohydrin and Poly-ethylene oxide

KNOWLEDGE GAP AND RESEARCH QUESTIONS

The authors of this article hold the view that despite the large volume of work done already in research within drag reduction and flow pattern modification using DRPs, several research questions still remain unanswered and there are still enormous opportunities for research within this dynamic field of expertise. Some of the identified areas of further research are highlighted in Sections 11.1 and 11.2.

Single Phase Flow

Drag reduction involving single phase turbulent flow has been deeply studied especially with regard to experimental and, to a much lesser extent, theoretical investigations. However, the underlying mechanism of the drag reduction phenomenon such as the interaction between the polymer additive and the turbulent eddy motion is still poorly understood and therefore remains the subject of much study. In spite of the many significant advances in the understanding – with the help of direct numerical simulations (DNS) – of the polymer drag reduction mechanism, maximum drag reduction and coherent structures, at least one fundamental question still remains to be answered: How does one predict the magnitude of drag reduction from the properties of the flow and polymer solution?

Multiphase Flow

Compared to the single phase flow, limited experimental works are available in multiphase flows most of which involve gas–liquid flows with liquid–liquid flows just recently gaining some attentions. Suffice it to say that virtually no theoretical models, either phenomenological or using DNS, are available in open literature till date to predict quantitatively the extent of drag reduction in multiphase flows, though a handful of empirical and mechanistic correlations can be found. In addition, the very few mechanisms of how drag reduction occurs are mere speculations. With this limited knowledge on the use of DRPs in multiphase flows, several research areas listed below still remain a concern and a lot of works is needed to be done to bridge this gap. Hence, exploitation of the enormous opportunities for research within this dynamic field of expertise cannot be over emphasized.

- Polymer degradation in multiphase flows. Solving or finding a solution to the problem of polymer degradation will be helpful in enhancing the effectiveness of DRPs.

- Further experimental investigations into the effect of DRPs, especially in liquid–liquid flows are still required in the areas of maximum drag reduction, new flow pattern maps, oil–water interfacial shapes and drop sizes.

- Drag reduction in vertical or near-vertical gas–liquid flows should be further investigated.

- From a practical or applied point of view, investigations of drag reduction in larger diameter of vertical or horizontal pipes may be rewarding.

- Pipe diameter scale-up correlation models are equally needed for industrial purposes.

- Some of the parameters affecting the performance of DRPs like liquid salinity and temperature have not been adequately investigated especially in oil–water flows. Therefore, more works need to be done in this area to address these effects.

- It is high time researchers focused on the use of high viscosity oil on the study of drag reduction in multiphase flows since the viscosity of the currently used oils are far lower than those of crude oil.

CONCLUSIONS

The construction of long distance pipelines crossing international borders for petroleum products transportation is increasing amidst the growing energy demand globally and so the need for drag reduction to minimize pumping cost and reduce separation time. This means that the research area of drag reduction will continue to receive attention even as more demands to challenge drag, pressure loss and the general understanding of turbulence in flows will be placed on it. The availability of detailed turbulence measurements coupled with DNS/molecular models has shed more light on the subject. However, there are still many areas of uncertainties to be addressed. Specific among these are tackling a major shortcoming of DRPs as drag reducers which is degradation and the development of more environmentally friendly and cheaper commercial DRPs. Little has been done on the effect of DRPs on liquid–liquid, gas–liquid–liquid and the authors hope research in these areas will be stretched farther.

ACKNOWLEDGMENTS

The authors would like to thank the Research Council (Oman) for providing financial support for the project.

REFERENCES

1.   Abid-Ali, Q.M., Al-ausi, T.A., 2008. Drag force reduction of flowing crude oil by polymers addition. Iraqi Journal for Mechanical and Material Engineering 8 (2), 149–161.

2.   Absi, F.S., Oliveira, T.F., Cunha, F.R., 2006. A note on the extensional viscosity of elastic liquids under strong flows. Mechanics Research Communications 33, 401–414.

3.   Al-Sarkhi, A., 2010. Drag reduction with polymers in gas–liquid/liquid–liquid flows in pipes: a literature review. Journal of Natural Gas Science and Engineering 2, 41–48.

4.   Al-Sarkhi, A., 2012. Effect of mixing on frictional loss reduction by drag reducing polymer in annular horizontal two-phase flows. International Journal of Multiphase Flow 39, 186–192.

5.   Al-Sarkhi, A., Abu-Nada, E., 2005. Effect of drag reducing polymer on annular flow patterns of air and water in a small horizontal pipeline. In: Twelfth International Conference on Multiphase Production Technology, Barcelona, Spain.

6.   Al-Sarkhi, A., Abu-Nada, E., Batayneh, M., 2006. Effect of drag reducing polymer on air–water annular flow in an inclined pipe. International Journal of Multiphase Flow 32, 926–934.

7.   Al-Sarkhi, A., El Nakla, M., Ahmed, W.H., 2011. Friction factor correlations for gas–liquid/liquid–liquid flows with drag-reducing polymers in horizontal pipes. International Journal of Multiphase Flow 37, 501–506.

8.   Al-Sarkhi, A., Hanratty, T.J., 2001a. Effect of drag-reducing polymers on annular gas–liquid flow in a horizontal pipe. International Journal of Multiphase Flow 27, 1151–1162.

9.   Al-Sarkhi, A., Hanratty, T.J., 2001b. Effect of pipe diameter on the performance of drag-reducing polymers in annular gas–liquid flows. Trans IChemE, Institution of Chemical Engineers 79 (Part A), 402–408.

10.   Al-Sarkhi, A., Soleimani, A., 2004. Effect of drag reducing polymers on two-phase gas–liquid flows in a horizontal pipe. Trans IChemE, Part A, Chemical Engineering Research and Design 82 (A12), 1583–1588.

11.   Al-Wahaibi, T., Al-Wahaibi, Y., Al-Ajmi, A., Yusuf, N., Al-Hashmi, A.R., Olawale, A.S., Mohammed, I.A., 2013. Experimental Investigation on the performance of drag reducing polymers through two pipe diameters in horizontal

oil–water flows. Experimental Thermal and Fluid Science, 1–25.

12. Al-Wahaibi, T., Smith, M., Angeli, P., 2007. Effect of drag-reducing polymers on horizontal oil–water flows. Journal of Petroleum Science and Engineering 57, 334–346.

13. Al-Yaari, M.A., (M.Sc. thesis) 2008. Influence of Drag Reducing Polymers on Oil–water Flow Characteristics. King Fahd University of Petroleum and Minerals, Dharhran, Saudi Arabia.

14. Al-Yaari, M., Al-Sarkhi, A., Abu-Sharkh, B., 2012. Effect of drag reducing polymers on water holdup in an oil–water horizontal flow. International Journal of Multiphase Flow 44, 29–33.

15. Al-Yaari, M., Soleimani, A., Abu-Sharkh, B., Al-Mubaiyedh, U., Al-sarkhi, A., 2009. Effect of drag reducing polymers on oil–water flow in a horizontal pipe. International Journal of

16. Multiphase Flow 35, 516–524.

17. Angeli, P., Hewitt, G.F., 2000. Flow structure in horizontal oil–water flow. International Journal of Multiphase Flow 26, 1117–1140.

18. Baik, S., Hanratty, T.J., 2003. Effects of a drag reducing polymer on stratified gas–liquid flow in a large diameter horizontal pipe. International Journal of Multiphase Flow 29, 1749–1757.

19. Baik, S., Vlachogannis, M., Hanratty, T., 2005. Use of particle image velocimetry to study heterogeneous drag reduction. Experiments in Fluids 39, 637–650.

20. Baron, A., Sibilla, S., 1997. Direct numerical simulation of turbulent channel flow of viscoelastic fluids. In: Geurts, B., Kuerten, H. (Eds.), DNS and LES of Complex Flows: Numerical and Modeling Aspects. Twente, Enschede, Netherlands, pp. 229–237.

21. Baron, A., Sibilla, S., 1998. DNS of the turbulent channel flow of a dilute polymer solution. Applied Scientific Research 59, 331–352.

22. Berman, N.S., 1978. Drag reduction by polymers. Annual Review of Fluid Mechanics 10, 47–64.

23. Bhattacharjee, J.K., Thirumalai, D., 1991. Drag reduction in turbulent flows by polymers. Physical Review Letters 67 (2), 196–199.

24. Bird, R.B., Armstrong, R.C., Hassager, O., 1987. Dynamics of Polymeric Liquids, 2nd ed. Wiley, New York.

25. Brostow, W., Ertepinar, H., Singh, R.P., 1990. Flow of dilute polymer solutions chain conformations and degradation of drag reducers. Macromolecule 23, 5109–5118.

26. Burger, E.D., Munk, W.R., Wahi, H.A., 1982. Flow increase in the trans-Alaska pipeline using a polymeric drag reducing additives. Journal of Petroleum Technology, 377–386.

27. Ceccio, S.L., Dowling, D.R., Perlin, M., Solomon, M., 2007. Influence of Surface Roughness on Polymer Drag Reduction. Final Technical Report of HROO11-06-1-0057. University of

28. Michigan, Ann Arbor.

29. Choi, H.J., Kim, C.A., Sohn, J., Jhon, M.S., 2000. An exponential decay function for polymer degradation in turbulent drag reduction. Polymer Degradation and Stability 69, 341–346.

30. Chung, J.S., Lee, K., Tischler, A., 2007. Two-phase vertically upward transport of silica sands in dilute polymer solution: drag reduction and effects of sand size and concentration. In:

31. Proceedings of The Seventh ISOPE Ocean Mining Symposium, Lisbon, Portugal, pp. 188–196.

32. Clifford, N.J., French, J.R., 1993. Monitoring and modeling turbulent flow: historical and contemporary. Turbulence: Perspectives on Flow & Sediment, 1–34.

33. Cruz, D.O.A., Pinho, F.T., Resende, P.R., 2004. Modeling the new stress for improved drag reduction predictions of viscoelastic pipe flow. Journal of Non-Newtonian Fluid Mechanics 121, 127–141.

34. Daas, M., Kang, C., Jepson, W.P., 2000. Quantitative analysis of drag reduction in horizontal slug flow. Society of Petroleum Engineers 62944, 1–8.

35. De Gennes, P.G., 1990. Introduction to Polymer Dynamics. University Cambridge, Cambridge. de Schepper, S.C.K., Heynderickx, G.J., Marin, G.B., 2008. CFD modeling of all gas–liquid and vapor–liquid flow regimes predicted by the Baker chart. Chemical Engineering Journal 138, 349–357.

36. Dembek, G., Bewersdorff, H.W., 1981. Short-time increase of sewer capacity by addition of water-soluble polymers. GWF Wasser/Abwasser 122 (9), 392–395.

37. den Toonder, J.M.J., Nieuwstadt, F.T.M., Kuiken, G.D.C., 1995. The role of elongational viscosity in the mechanism of drag reduction by polymer additives. Applied Sciences Research 54, 95–123.

38. Deshmukh, S.R., Sudhakar, K., Singh, R.P., 1991. Drag reduction efficiency, shear stability and biodegradability resistance of carboxymethyl cellulose based and starch based graft copolymers. Journal of Applied Polymer Science 43, 1091–1101.

39. Dimitropoulos, C.D., Sureshkumar, R., Beris, A.N., 1998. Direct numerical simulation of viscoelastic turbulent channel flow exhibiting drag reduction: effect of the variation of rheological parameters. Journal of Non-Newtonian Fluid Mechanics 79, 433–468.

40. Dubief, Y., Terrapon, V., White, C., Shaqfeh, E., Moin, P., Lele, S., 2005. New answers on the interaction between polymers and vortices in turbulent flows. Flow, Turbulence and Combustion 74 (4), 311–329.

41. Dubief, Y., White, C.M., Terrapon, V.E., Shaqfeh, E.S.G., Moin, P., Lele, S.K., 2004. On the coherent drag reducing and turbulence enhancing behaviour of polymers in wall flows. Journal of Fluid Mechanics 514, 271–280.

42. Edomwonyi-Otu, L., Barral, A., Angeli, P., 2013. Influence of drag reducing agents on interfacial wave characteristics in

horizontal oil–water flow. In: 16th International Conference on Multiphase Production Technology, Cannes, France 12th–14th June.

43. Eichelberger, D.P., (M.Sc. thesis) 1992. Molecular Interactions of Water-soluble Polymer Blends and their Effect on Drag Reduction in Dilute Aqueous Solutions. Lehigh University, Bethlehem, United States.

44. Elperin, I.T., Smolskii, B.M., Leventhal, L.I., 1967. Decreasing the hydrodynamic resistance of pipelines. International Journal of Chemical Engineering 7, 276–295.

45. Escudier, M.P., Poole, R.J., Presti, F., Dales, C., Nouar, C., Desaubry, C., Graham, L., Pullum, L., 2005. Observations of asymmetrical flow behaviour in transitional pipe flow of yield-stress and other shear-thinning liquids. Journal of Non-Newtonian Fluid Mechanics 127, 143–155.

46. Escudier, M.P., Presti, F., 1996. Pipe flow of a thixotropic liquid. Journal of Non-Newtonian Fluid Mechanics 62, 291–306.

47. Escudier, M.P., Rosa, S., Poole, R.J., 2009. Asymmetry in transitional pipe flow of drag-reducing polymer solutions. Journal of Non-Newtonian Fluid Mechanics 161, 19–29.

48. Fernandes, R.L.J., Jutte, B.M., Rodriguez, M.G., 2004. Drag reduction in horizontal annular two-phase flow. International Journal of Multiphase Flow 30, 1051–1069.

49. Fox, K.B., Bainum, M.S., Chemical, C.,2010. A new synthetic polymer provides improved drag reduction in coiled tubing operations in North Louisiana. In: SPE/ICoTA coiled tubing and well intervention conference and exhibition. TX, USA, pp. 1–8.

50. Gaard, S., Isaksen, O.T., 2003. Experiments with Various Drag Reducing Additives in Turbulent Flow in Dense Phase Gas Pipelines. Pipeline Simulation Interest Group, Bern, Switzerland, pp. 1–12.

51. Gallego, F., Shah, S.N., 2009. Friction pressure correlations for turbulent flow of drag reducing polymer solutions in

straight and coiled tubing. Journal of Petroleum Science and Engineering 65, 147–161.

52.   Gasljevic, K., Aguilar, G., Matthys, E.F., 2001. On two distinct types of drag-reducing fluids, diameter scaling, and turbulent profiles. Journal of Non-Newtonian Fluid Mechanics 96, 405–425.

53.   Greskovich, E.J., Shrier, A.L., 1971. Drag reduction in two-phase flows. Industrial & Engineering Chemistry Fundamentals 10, 646–648.

54.   Gyr, A., Bewersdorff, H.W., 1995. Drag Reduction of Turbulent Flows by Additives. Kluwer Academic, Dordrecht, Netherlands.

55.   Hadri, F., Besq, A., Guillou, S., Makhlouf, R., 2011. Temperature and concentration influence on drag reduction of very low concentrated CTAC/nasal aqueous solution in turbulent pipe flow. Journal of Non-Newtonian Fluid Mechanics 166, 326–331.

56.   Handler, R.A., Swean, T.F., Leighton, R.I., Swearingen, J.D., 1993. Length scales and the energy-balance for turbulence near a free surface. AIAA Journal 31, 1998–2007.

57.   Harder, K.J., Tiederman, W.G., 1991. Drag reduction and turbulent structure in two-dimensional channel flow. Philosophical Society of the Royal Society, Series A 336, 19–28.

58.   Henaut, I., Darbouret, M., Palermo, T., Glenat, P., Hurtevent, C., 2009. Experimental methodology to evaluate drag reducing agents: effect of water content and waxes on their efficiency. In: International Symposium, Society of Petroleum Engineers, TX, USA.

59.   Hénaut, I., Glénat, P., Cassar, C., Gainville, M., Hamdi, K., Pagnier, P., 2012. Mechanical Degradation Kinetics of Polymeric DRAs. IFP Energies Nouvelles, France.

60.   Hershey, H.C., Zakin, J.L., 1967. Molecular approach to predicting the onset of drag reduction in the turbulent flow of

dilute polymer solutions. Chemical Engineering Science 22, 1847–1857.

61. Hibberd, M., Kwade, M., Scharf, R., 1982. Influence of drag reducing additives on the structure of turbulence in a mixing layer. Rheologica Acta 21, 582–586.

62. Hou, Y.X., Somandepalli, V.S.R., Mungal, M.G., 2006. A technique to determine total shear stress and polymer stress profiles in drag reduced boundary layer flows. Experiments in Fluids 40 (4), 589–600.

63. Hoyt, J.W., 1980. Effect of ferric ions on drag reduction effectiveness of polyacrylamide. Polymer Engineering & Science 20 (7), 493–498.

64. Hoyt, J.W., 1989. In: Busnell, D.M., Hefner, J.M. (Eds.), Drag reduction by Polymers And Surfactants, Viscous Drag Reduction in Boundary Layers. American Institute of Aeronautics and Astronautics. Iaccarino, G., Shaqfeh, E.S.G., Dubief, Y., 2010. Reynolds-averaged modeling of polymer drag reduction in turbulent flows. Journal of Non-Newtonian Fluid Mechanics 165, 376–384.

65. Ilg, P., De Angelis, E., Karlin, I.V., Casciola, C.M., Succi, S., 2002. Polymer dynamics in wall turbulent flow. Europhysics Letters 58, 616–622.

66. Interthal, W., Wilski, H., 1985. Drag reduction experiments with very large pipes. Colloid and Polymer Science 263, 217–229.

67. Ioannou, K., Nydal, O.J., Angeli, P., 2005. Phase inversion in dispersed liquid–liquid flows. Experimental Thermal and Fluid Science 29 (3), 331–339.

68. Japper-Jaafar, A., Escudier, M.B., Poole, R.J., 2010. Laminar, transitional and turbulent annular flow of drag-reducing polymer solutions. Journal of Non-Newtonian Fluid

69. Mechanics 165, 1357–1372.

70. Japper-Jaafar, A., Escudier, M.P., Poole, R.J., 2009. Turbulent pipe flow of a drag-reducing rigid rod-like polymer solution. Journal of Non-Newtonian Fluid Mechanics 161, 86–93.

71. Jensen, K.D., 2004. Flow measurements. Journal of the Brazilian Society of Mechanical Sciences and Engineering XXVI (4), 400–419.

72. Jha, B.K., Chauhan, V.S., Sharma, S., Khanna, A.S., 2012a. Potential energy conservation technique for fluid transportation through pipeline: water soluble polymer drag reduction

73. technology. In: IPCBEE (Ed.), 2nd International Conference on Environment and Industrial Innovation. Singapore, pp. 103–107.

74. Jha, B.K., Yanjarappa, M.J., Khanna, A.S., 2012b. Strategic polymer drag reduction technology for energy conservation: modification of injection system for fluid transportation through pipeline. International Journal of Chemical and Petrochemical Technology 2 (3), 1–8.

75. Jia, N., Gourma, M., Thompson, C.P., 2011. Non-Newtonian multi-phase flows: on drag reduction, pressure drop and liquid wall friction factor. Chemical Engineering Science 66,

76. 4742–4756.

77. Jourdan, L., Knapp, Y., Oliver, F., Guibergia, J.P., 1998. The effect of drag-reducing polymer additives on wall-pressure fluctuations in turbulent channel flows. European Journal of

78. Mechanics, B/Fluid 17 (1), 105–136.

79. Jubran, B.A., Zurigat, Y.H., Goosen, M.F.A., 2005. Drag reducing agents in multiphase flow pipelines: recent trends and future needs. Petroleum Science and Technology 23, 1403–1424.

80. Kamel, A.H.A., 2011. Drag reduction behavior of polymers in straight and coiled tubing at elevated temperature. Electronic Scientific Journal of Oil and Gas Business 1 (1), 107–128.

81. Kameneva, M.V., Wu, Z.J., Uraysh, A., 2004. Blood soluble drag-reducing polymers prevent lethality from hemorrhagic shock in acute animal experiments. Biorheology 41 (1), 53–64.

82. Kang, C., Jepson, W.P., 1999. Multiphase flow conditioning using drag-reducing agents. Society of Petroleum Engineers 56569, 1–7.

83. Kang, C., Jepson, W.P., 2000. Effect of drag-reducing agents in multiphase, oil/gas horizontal flow. Society of Petroleum Engineers 58976, 1–7.

84. Kang, C., Kerr, H., Vancko, R.M., Green, A.S., Jepson, W.P., 1998. Effect of drag-reducing agents in multiphase flow pipelines. Journal of Energy Resources Technology 120, 15–19.

85. Karami, H.R., Mowla, D., 2012. Investigation of the effects of various parameters on pressure drop reduction in crude oil pipelines by drag reducing agents. Journal of Non-Newtonian Fluid Mechanics 177–178, 37–45.

86. Kim, C.A., Kim, J.T., Lee, K., Choia, H.J., Jhon, M.S., 2000. Mechanical degradation of dilute polymer solutions under turbulent flow. Polymer 41, 7611–7615.

87. Kim, K., Islam, M.T., Shen, X., Sirviente, A.I., Solomon, M.J., 2004. Effect of macromolecular polymer structures on drag reduction in a turbulent channel flow. Physics of Fluids 16 (11), 4150–4162.

88. Kim, K., Sirviente, A.I., 2005. Turbulence structure of polymer turbulent channel flow with and without macromolecular polymer structures. Experiments Fluids 38, 739–749.

89. Kim, N., Kim, S., Lim, S.H., Chen, K., Chun, W., 2009.

90. Measurement of drag reduction in polymer added turbulent flow. International Communications in Heat and Mass Transfer 36, 1014–1019.

91. Kumar, S., Homsy, G.M., 1999. Direct numerical simulation of hydrodynamic instabilities in two- and three-dimensional viscoelastic free shear layers. Journal of Non-Newtonian Fluid Mechanics 83, 249–276.

92. Landahl, M.T., 1977. Dynamics of boundary layer turbulence and the mechanism of drag reduction. Physics of Fluids 20 (10), S55–S63.

93. Langsholt, M., 2012. An Experimental Study on Polymeric Type DRA used in Single- and Multiphase Flow with Emphasis on Degradation, Diameter Scaling and the Effects in Three-phase Oil–Water–Gas Flow. Institute for Energy Technology (IFE), Norway.

94. Larson, R.G., 2003. Analysis of polymer turbulent drag reduction in flow past a flat plate. Journal of Non-Newtonian Fluid Mechanics 111, 229–250.

95. Launder, B.E., Sharma, B.I., 1974. Application of the energy-dissipation model of turbulence to the calculation of flow near a spinning disc. Letters in Heat and Mass Transfer 1 (2), 131–138.

96. Lee, K., Kim, C.A., Lim, S.T., Kwon, D.H., Choi, H.J., Jhon, M.S., 2002. Mechanical degradation of polyisobutylene under turbulent flow. Colloid and Polymer Science 280, 779–782.

97. Lescarboura, J.A., Culter, J.D., Wahl, H.A., 1971. Drag reduction with a polymeric additive in crude oil pipelines. SPE Journal 11 (3), 229–235.

98. Liaw, G.C., (Ph.D. Dissertation) 1969. Effect of Polymer Structure on Drag Reduction in Nonpolar Solvents. University of Missouri-Rolla, USA.

99. Liaw, G.C., Zakin, J.L., Patterson, G.K., 1971. Effects of molecular characteristics of polymers on drag reduction. AIChE Journal 17, 391–397.

100. Liberatore, M.W., Baik, S., McHugh, A.J., Hanratty, T.J., 2004. Turbulent drag reduction of polyacrylamide solutions: effect of degradation on molecular weight distribution. Journal of Non-Newtonian Fluid Mechanics 123, 175–183.

101. Lumley, J.L., 1969. Drag reduction by additives. Annual Review of

102. Fluid Mechanics 1, 367–384.

103. Lumley, J.L., 1973. Drag reduction in turbulent flow by polymer additives. Macromolecular Reviews 7 (1), 263–290.

104. Lumley, J.L., 1977. Drag reduction in two phase and polymer flows. Physics of Fluids 20, S64–S71.

105. L'vov, V.S., Pomyalov, A., Procaccia, I., Tiberkevich, V., 2004. Drag reduction by polymers in wall bounded turbulence. Physical Review Letters 92 (24), 244503(1)–244503(4).

106. L'vov, V.S., Pomyalov, A., Procaccia, I., Tiberkevich, V., 2005. The polymer stress tensor in turbulent shear flow. Physical Review E 71, CD/0405022.

107. Manfield, P.D., Lawrence, C.J., Hewith, G.F., 1999. Drag reduction with additives in multiphase flow: a literature survey. Multiphase Science and Technology 11, 17–221.

108. Marmy, R.M.S., Hayder, A.B., Rosli, M.Y., 2012. Improving the flow in pipelines by Cocos nucifera fiber waste. International Journal of Physical Sciences 7 (26), 4073–4080.

109. Martin, J.R., Shapella, B.D., 2003. The effect of solvent solubility parameter on turbulent flow drag reduction in polyisobutylene solutions. Experiments in Fluids 34, 535–539.

110. Massah, H., Hanratty, T.J., 1997. Added stresses because of the presence of FENE-P bead-spring chains in a random velocity field. Journal of Fluid Mechanics 337, 67–101.

111. Massah, H., Kontomaris, K., Schowalter, W.R., Hanratty, T.J., 1993. The configurations of a FENE bead-spring chain in transient rheological flows and in a turbulent flow. Physics of Fluids A 5 (4), 881–890.

112. McComb, W.D., Rabie, L.H., 1979. Development of local turbulent drag reduction due to non-uniform polymer concentration. Physics of Fluids 22, 183–186.

113. McComb, W.D., Rabie, L.H., 1982. Local drag reduction due to injection of polymer solutions into turbulent flow in a pipe. AIChE Journal 28, 547–565.

114. Mehrabadi, M.A., Sadeghy, K., 2008. Simulating drag reduction phenomenon in turbulent pipe flows. Mechanics Research Communications 35, 609–613.

115. Min, T., Yoo, J.Y., Choi, H., 2001. Effect of spatial discretization schemes on numerical solutions of viscoelastic fluid flows. Journal of Non-Newtonian Fluid Mechanics 100, 27–47.

116. Mohsenipour, A.A., Pal, R., 2013. Drag reduction in turbulent pipeline flow of mixed nonionic polymer and cationic surfactant systems. Canadian Journal of Chemical

117. Engineering 91, 190–201.

118. Mohsenipour, A.A., Pal, R., Prajapati, K., 2013. Effect of cationic surfactant addition on the drag reduction behaviour of anionic polymer solutions. Canadian Journal of Chemical Engineering 91, 181–189.

119. Mostardi, R.A., Thomas, L.C., Greene, H.L., VanEssen, F., Nokes, R.F., 1978. Suppression of atherosclerosis in rabbits using drag reducing polymers. Biorheology 15 (1), 1–14.

120. Motozawa, M., Ishitsuka, S., Iwamoto, K., Ando, H., Senda, T., Kawaguchi, Y., 2012. Experimental investigation on turbulent structure of drag reducing channel flow with blowing polymer solution from the wall. Flow, Turbulence and Combustion 88, 121–141.

121. Moussa, T., Tiu, C., 1994. Factors affecting polymer degradation in turbulent pipe flow. Chemical Engineering Science 49, 1681–1692.

122. Mowla, D., Naderi, A., 2006. Experimental study of drag reduction by a polymeric additive in slug two-phase flow of crude oil and air in horizontal pipes. Chemical Engineering Science 61 (5), 1549–1554.

123. Nakken, T., Tande, M., Elgsaeter, A., 2001. Measurements of polymer induced drag reduction and polymer scission in Taylor flow using standard double-gap sample holders with axial symmetry. Journal of Non-Newtonian Fluid Mechanics 97, 1–12.

124. Nesyn, G.V., Manzhai, V.N., Suleimanova, Y.V., Stankevich, V.S., Konovalov, K.B., 2012. Polymer drag reducing agents for transportation of hydrocarbon liquids: mechanism of action, estimation of efficiency, and features of production. Polymer Science, Series A 54 (1), 61–67.

125. Oldroyd, J.G., 1949. A suggested method of detecting wall-effect in turbulent flow through tubes. In: Proc. 1st Intern. Congr. Rheol. North Holland, pp. II130–II134.

126. Oliver, D.R., Young, H.A., 1968. Two-phase non-Newtonian flow. Transactions of the Institution of Chemical Engineers 46, T106.

127. Omer, A., Pal, R., 2010. Pipeline flow behavior of water-in-oil emulsions with and without a polymeric additive. Chemical Engineering Technology 33 (6), 983–992.

128. Otten, L., Fayed, A.S., 1976. Pressure drop and drag reduction in two-phase non-Newtonian slug flow. Canadian Journal of Chemical Engineering 54, 111–114.

129. Parimal, P., Cheolho, K., Alvaro, A., 2008. The performance of drag reducing agents in multiphase flow conditions at high pressure: positive and negative effects. In: IPC2008, Seventh International Pipeline Conference, Calgary, Alberta, Canada.

130. Paschkewitz, J.S., Dimitropoulos, C.D., Hou, Y.X., Somandepalli, V.S.R., Mungal, M.G., Shaqfeh, E.S.G., Moin, P., 2005. An experimental and numerical investigation of drag reduction in a turbulent boundary layer using a rigid rod-like polymer. Physics of Fluids 17, 1–18.

131. Peixinho, J., Nouar, C., Desaubry, C., Théron, B., 2005. Laminar transitional and turbulent flow of yield stress fluid in a pipe. Journal of Non-Newtonian Fluid Mechanics 128, 172–184.

132. Pereira, A.S., Soares, E.J., 2012. Polymer degradation of dilute solutions in turbulent drag reducing flows in a cylindrical double gap rheometer device. Journal of Non-Newtonian Fluid Mechanics 179–180, 9–22.

133. Peterlin, A., 1966. Hydrodynamics of macromolecules in a velocity field with longitudinal gradient. Journal of Polymer Science Part C: Polymer Letters 4B, 287–291.

134. Petrie, H.L., Deutsch, S., Brungart, T.A., Fontaine, A.A., 2003.

135. Polymer drag reduction with surface roughness in flat-plate turbulent boundary layer flow. Experiments in Fluids 35, 8–23.

136. Pinho, F.T., 2003. A GNF framework for turbulent flow models of drag reducing fluids and proposal for a k–ε type closure. Journal of Non-Newtonian Fluid Mechanics 114, 149–184.

137. Pinho, F.T., Sadanandan, B., Sureshkumar, R., 2008. One equation model for turbulent channel flow with second order viscoelastic corrections. Flow, Turbulence and Combustion 81, 337–367.

138. Pinho, F.T., Whitelaw, J.H., 1990. Flow of non-newtonian fluids in a pipe. Journal of Non-Newtonian Fluid Mechanics 34 (2), 129–144.

139. Ptasinski, P.K., Boersma, B.J., Nieuwstadt, F.T.M., Hulsen, M.A., Van Den Brule, B.H.A.A., Hunt, J.C.R., 2003. Turbulent channel flow near maximum drag reduction: simulations, experiments and mechanism. Journal of Fluid Mechanics 490, 251–291.

140. Ptasinski, P.K., Nieuwstadt, F.T.M., Van-Den, B.H.A.A., Hulsen, M.A., 2001. Experiments in turbulent pipe flow with polymer additives at maximum drag reduction. Turbulence Combustion 66, 159–182.

141. Radin, I., Zakin, J.L., Patterson, J.K., 1975. Drag reduction in solid–fluid systems. AIChE Journal 21 (2), 358–371.

142. Resende, P.R., Escudier, M.P., Presti, F., Pinho, F.T., Cruz, D.O.A., 2006. Numerical predictions and measurements of Reynolds normal stresses in turbulent pipe flow of polymers. International Journal of Heat and Fluid Flow 27, 204–219.

143. Resende, P.R., Kim, K., Younis, B.A., Sureshkumar, R., Pinho, F.T., 2011. A FENE-P k–ε turbulence model for low and intermediate regimes of polymer-induced drag reduction. Journal of Non-Newtonian Fluid Mechanics 166, 639–660.

144. Rosehart, R.G., Scott, D., Rhodes, E., 1972. Gas–liquid slug flow with drag-reducing polymer solutions. AIChE Journal 18 (4), 744–750.

145. Ryskin, G., 1987. Turbulent drag reduction by polymers: a quantitative theory. Physical Review Letters 59 (18), 2059–2062.

146. Saether, G., Kubberud, K., Nuland, S., Lingelem, M.N., 1989. Drag reduction in two phase flow. In: Fourth International Conference on Multiphase Flow, France, pp. 171–184.

147. Schadel, S.A., Leman, G.W., Binder, J.L., Hanratty, T.J., 1990. Rates of atomization and deposition in vertical annular flow. International Journal of Multiphase Flow 16, 363–374.

148. Scott, D., Rhodes, E., 1972. Gas–liquid slug flow with drag reducing polymer solutions. AIChE Journal 18, 744–750.

149. Shah, S.N., Zhou, Y., 2003. An Experimental study of drag reduction of polymer solutions in coiled tubing. SPE Production & Facilities, 280–287.

150. Shah, S.N., Zhou, Y., 2009. Maximum drag reduction asymptote of polymeric fluid flow in coiled tubing. Journal of Fluids Engineering (ASME) 131, 011201(1)–011201(9).

151. Shanshool, J., Al-Qamaje, H.M.T., 2008. Effect of molecular weight on turbulent drag reduction with polyisobutylene. NUCEJ Spatial 11 (1), 52–59.

152. Shanshool, J., Marwa, F., Jabbar, A., Sulaiman, I.N., 2011. The influence of mechanical effects on degradation of polyisobutylenes as drag reducing agents. Petroleum & Coal 53 (3), 218–222.

153. Sher, I., Hetsroni, G., 2008. A mechanistic model of turbulent drag reduction by additives. Chemical Engineering Science 63, 1771–1778.

154. Sifferman, T.R., Greenkorn, R.A., 1981. Drag reduction in three distinctly different fluid systems. Society of Petroleum Engineers Journal, 663–669.

155. Singh, R.P., Jai, S.K., Lan, N., 1991. In: Sivaram, S. (Ed.), Drag Reduction, Flocculation and Rheological Characteristics of Grafted Polysaccharides. Polymer Science Contemporary

156. Themes, New Delhi, p. 716.

157. Smith, R.E., Tiederman, W.G., 1991. The mechanism of polymer thread drag reduction. Rheological Acta 30, 103–113.
158. Soleimani, A., Al-Sarkhi, A., Hanratty, T.J., 2002. Effect of drag-reducing polymers on pseudo-slugs—interfacial drag and transition to slug flow. International Journal of Multiphase Flow 28, 1911–1927.
159. Sood, A., Rhodes, W., 1998. Pipeline scale-up in drag reducing turbulent flow. Canadian Journal of Chemical Engineering 76, 11–18.
160. Sreenivasan, K.R., White, C.M., 2000. The onset of drag reduction by dilute polymer additives, and the maximum drag reduction asymptote. Journal of Fluid Mechanics 409, 149–164.
161. Sureshkumar, R., Beris, A.N., Handler, R.A., 1997. Direct numerical simulation of the turbulent channel flow of a polymer solution. Physics of Fluids 9, 743–755.
162. Sylvester, N.D., Brill, J.P., 1976. Drag-reduction in two-phase annular mist flow of air and water. AIChE Journal 22 (3), 615–617.
163. Sylvester, N.D., Dowling, R.H., Brill, J.P., 1980. Drag reductions in concurrent horizontal natural gas–hexane pipe flow. Polymer Engineering & Science 20, 485.
164. Toms, B.A., 1948. Some observations on the flow of linear polymer solutions through straight tubes at large Reynolds numbers. In: First International Congress on Rheology, vol. 2, Amsterdam, pp. 135–141.
165. Toonder, J.M.J., (Ph.D. thesis) 1995. Drag Reduction by Polymer Additives in a Turbulent Pipe Flow: Laboratory and Numerical Experiments. Delft University of Technology, Delft, South Holland, Netherlands.
166. Truong, V.T., 2001. Drag Reduction Technologies. Maritime Platforms Division, Australia.

167. Tulin, M., 1984. In: Sellin, R.H.J., Moses, R.T. (Eds.), 3rd International Conference on Drag Reduction. University of Bristol, Bristol, UK.

168. Unthank, J.L., Lalka, S.G., Nixon, J.C., Sawchuk, A.P., 1992. Improvement of flow through arterial stenoses by drag reducing agents. Journal of Surgical Research 53 (6), 625–630.

169. Vaithianathan, T., Collins, L.R., 2003. Numerical approach to simulating turbulent flow of a viscoelastic polymer solution. Journal of Computational Physics 187, 1–21.

170. Vanapalli, S.A., Islam, M.T., Solomon, M.J., 2005. Scission-induced bounds on maximum polymer drag reduction in turbulent flow. Physics of Fluids 17, 095–108.

171. Virk, P.S., 1975. Drag reduction fundamentals. AIChE Journal 21, 625–656.

172. Virk, P.S., 1971. Drag reduction in rough pipes. Journal of Fluid Mechanics 45 (2), 225–246.

173. Virk, P.S., Merill, E.W., 1969. In: Wells, C.S. (Ed.), The Onset of Dilute Polymer Solution Phenomena in Viscous Drag Reduction: Viscous Drag Reduction. Plenum Press, New York.

174. Virk, P.S., Merrill, E.W., Mickley, H.S., Smith, K.A., Mollo-Christensen, E.L., 1967. The Toms phenomenon: turbulent pipe flow of dilute polymer solutions. Journal of Fluid Mechanic s 30, 305–328.

175. Virk, P.S., Wagger, D.L., 1989. Aspects of mechanisms in Type B drag reduction. In: Gyr, A. (Ed.), Structure of Turbulence and Drag Reduction. Springer-Verlag, Berlin, pp. 201–213.

176. Vlachogiannis, M., Hanratty, T.J., 2004. Influence of wavy structured surfaces and large scale polymer structures on drag reduction. Experiments in Fluids 36, 685–700.

177. Vlachogiannis, M., Liberatore, M.W., McHugh, A.J., Hanratty, T.J., 2003. Effectiveness of a drag reducing polymer: relation to molecular weight distribution and structuring. Physics of Fluids 15, 3786–3794.

178. Von Karman, T., 1930. Mechanical similitude and turbulence. Tech. Mem. NACA, no. 611.

179. Wallis, G.B., 1969. One-dimensional Two-Phase Flow. McGraw-Hill, New York.

180. Wang, Y., Yu, B., Zakin, J.L., Shi, H., 2011. Review on drag reduction and its heat transfer by additives. Advances in Mechanical Engineering, 1–17.

181. Warholic, M.D., Heist, D.K., Katcher, M., Hanratty, T.J., 2001. A study with particle-image velocimetry of the influence of drag-reducing polymers on the structure of turbulence. Experiment in Fluids 31, 474–483.

182. Warholic, M.D., Massah, H., Hanratty, T.J., 1999. Influence of drag-reducing polymers on turbulence: effects of Reynolds number, concentration and mixing. Experiment in Fluids 27, 461–472.

183. White, C.M., Mungal, M.G., 2008. Mechanics and prediction of turbulent drag reduction with polymer additives. Annual Review of Fluid Mechanics 40, 235–256.

184. White, C.M., Somandepalli, V.S.R., Mungal, M.G., 2004. The turbulence structure of drag-reduced boundary layer flow. Experiment in Fluids 36, 62–69.

185. Wilkens, R.J., Thomas, D.K., 2007. Multiphase drag reduction: effect of eliminating slugs. International Journal of Multiphase Flow 33, 134–146.

186. Wyatt, N.B., Gunther, C.M., Liberatore, M.W., 2011. Drag reduction effectiveness of dilute and entangled xanthan in turbulent pipe flow. Journal of Non-Newtonian Fluid Mechanics 166, 25–31.

187. Xu, X-X., 2007. Study on oil–water two-phase flow in horizontal pipelines. Journal of Petroleum Science and Engineering 59, 43–58.

188. Xu, J., Wu, Y., Li, H., Guo, J., Chang, Y., 2009. Study of drag reduction by gas injection for power-law fluid flow

in horizontal stratified and slug flow regimes. Chemical Engineering Journal 147, 235–244.

189. Yeh, Y., Cummins, H.Z., 1964. Localized fluid flow measurements with a He–Ne laser spectrometer. Applied Physics Letters 4, 176.

190. Yusuf, N., Al-Wahaibi, T., Al-Wahaibi, Y., Al-Ajmi, A., Al-Hahmi, A.R., Olawale, A.S., Mohammed, I.A., 2012. Experimental study on the effect of drag reducing polymer on flow patterns and drag reduction in a horizontal oil–water flow. International

191. Journal of Heat and Fluid Flow 37, 74–80.

192. Zadrazil, I., Bismarck, A., Hewitt, G.F., Markides, C.N., 2012. Shear layers in the turbulent pipe flow of drag reducing polymer solutions. Chemical Engineering Science 72, 142–152.

193. Zeybek, S., (M.Sc. thesis) 2005. Experimental Investigation of Drag Reduction Effects of Polymer Additives on Turbulent Pipe Flow. Middle East Technical University, Ankara, Turkey.

194. Zhou, Q., Akhavan, R., 2003. A comparison of FENE and FENE-P dumbbell and chain models in turbulent flow. Journal of Non-Newtonian Fluid Mechanics 109, 115–155.

195. Zhou, Y., Shah, S.N., Gujar, P.V., 2006. Effect of coiled-tubing curvature on drag reduction of polymeric fluids. Society of Petroleum Engineers 89478, 134–141.

Chapter 5

Analysis of Hydrodynamics of Fluid Flow on Corrugated Sheets of Packing

Kumar Subramanian and Günter Wozny

Berlin Institute of Technology, Straß des 17 Juni 135 Berlin, Germany

ABSTRACT

Modelling of the hydrodynamics behaviour of the liquid on the corrugated sheets of packing is studied using three-dimensional, volume-of-fluid (VOF) model that is incorporated in Ansys Fluent 12.0. The flow of three different liquids with different physical properties is modelled. A domain of corrugated sheets of packing resembling the real structured packing with little modifications in the elementary geometry is constructed using ICEM CFD 12.0. The quantitative comparisons of the wetting behavior from the simulations are in good agreement with experiments. Further, the study has been extended to understand the influence of the second corrugated sheet on the flow behavior. The contours from the simulations indicate the liquid hold-up in the crimp of two corrugated sheets, and these results are in good agreement with the earlier experimental studies performed using X-ray tomography in the literature. The result from the simulation shows that even for the high flow rate of around 811 mL/min for silicon-oil (DC5), only 60% of the corrugated sheet has been wetted. Hence, the efficiency of the existing packing can be further increased by increasing the wetted area in the corrugated sheet of packing.

INTRODUCTION

The key to success in separation of liquid mixtures by distillation depends on the creation and utilization of vapor-liquid contact area. The three major types of distillation equipments are trays, random packing, and structured packing. The corrugated sheet of packing, introduced in the late 1970s, became competent by claiming higher capacity and lower sensitivity to solids while retaining the high efficiency. Hence, by 1980s, the corrugated sheets of structured packing have drawn increased attention in industry [1]. The reason for the increased efficiency with structured packing is reported as additional mass transfer due to the increase in interfacial area created by spreading liquid over the packing surface [1]. Even though the structured packing is well established, the local flow behavior inside the packing is still not well understood. Various efforts have been exerted by researchers around the world using different approaches to know about the local flow behavior that helps to design the packed column and in terms of increasing their efficiency.

Geometrical features of corrugated sheets of packing play a crucial role in flow behavior of fluid inside the packing. Geometrical modifications of the packing are possible in three different ways: (1) varying surface of the corrugated sheet with grooved, lanced, textured, or smoothened surface, (2) changing the size of small elemental geometry like corrugation size and angle, and (3) with or without perforations.

Elementary Geometry

An elementary geometry of the corrugated sheet of packing is shown in Figure 1. The corrugation size defines the opening between adjacent corrugated layers. The ratio of B to h and S to h and the crimp angle (β) define the geometry of flow channel and of the vapor-liquid contact zone, respectively. Packing can be classified based on the specific surface area. Crimp angle varies from 28° to 45°, and base-to-height ratios range from 2 : 1 to 4 : 1. Most of the packing is not strictly triangle as shown below but is rounded top apex. The corrugation angle (α) also plays important role in deciding the capacity of the packing.

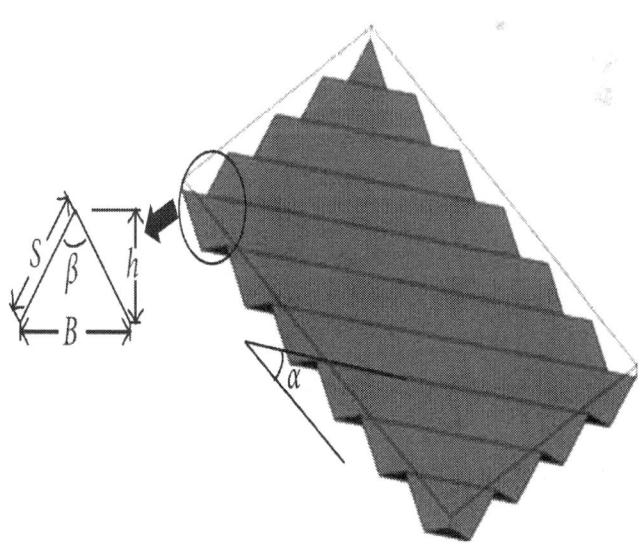

Figure 1: Elementary geometry details of the corrugated sheet of packing.

The element height is relatively low (typically 20 to 28 cm), and the angle of rotation is around 90°. Corrugation angle of 45° to the vertical enables good flow of liquid and avoids liquid accumulation.

The effect of the opening angle on the performance of the structured packing was studied earlier [2]. Two different opening angles of 90° and 20° were both studied numerical- and experimental-wise. It was also presented that when the opening angle was decreased from 90 to 20°, pressure drop of the packing could reduce by 35% and mass transfer could increase by 13% compared to Mellapak packing having the same specific surface area [2].

The results of two different packing series from Montz GmbH namely, B1 (embossed sheet metal, nonperforated) and BSH (expanded metal, perforated), were studied. In total, six different packings with two different corrugation angles of 45° and 60° and two different specific surfaces of 250 and 400 m²/m³ were investigated. The outcome showed that with increasing corrugation angle the pressure drop decreases, the capacity increases, and the mass transfer decreases. The holes on the surface could be the reason behind the slightly larger capacity of BSH packing in comparison with B1 packing. The influence of the corrugation angle on the performance of the packing was also presented [3].

Surface Features

Most of the structured packing surfaces have roughened or enhanced surfaces that assist the lateral spread of liquid, promote film turbulence, and enhance the area available for mass transfer. Measurements performed in the laboratory scale showed that mass transfer efficiency and wetter area are enhanced by textured surfaces. Texturing is of various types like grooving, lancing, shallow embossing, and deep embossing. Some different surface textures available in corrugated sheet of packing is shown in Figure 2. Figure 2(a) shows the corrugated sheet of packing without any surface textures but with perforations. Figures 2(b) and 2(c) show the three-dimensional surface textures on the packing. Various research studies have contributed exclusively to the study of the influence of surface textures on flow behavior and their further impact on mass transfer. The effect of 2-dimensional roughnesses on the flow behavior followed by gas-liquid absorption was reported [4]. The rate of absorption of CO_2 into water flowing over a plate with large-scale roughness can go up to 3.5 times faster than a smooth plate. The results were also compared with theoretical correlations. Similarly, various research contributions were presented to know the influence of film flow on complex and periodic surfaces [5–7]. The surface of most of the structured packing contains perforations.

(a)

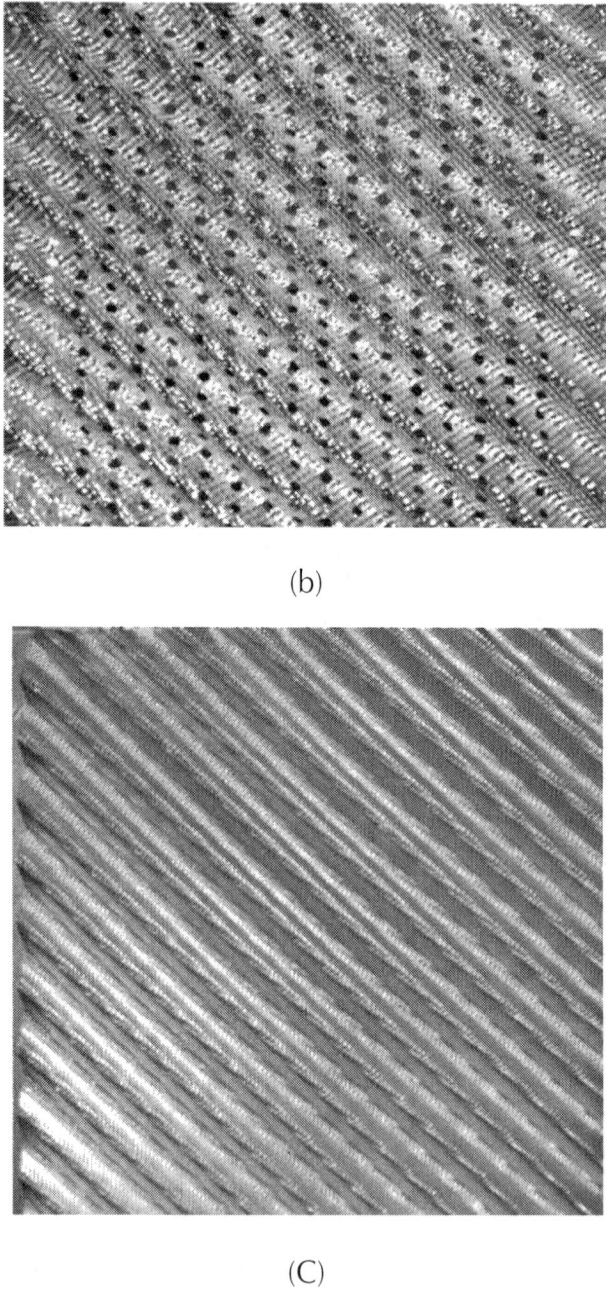

(b)

(C)

Figure 2: Different packings used in industrial applications: (a) smooth, perforated; (b) 3D grooved, perforated; (c) 3D embossed, unperforated.

CFD Studies

The geometry representing the column of 960 mm height with 288 mm inner diameter was meshed. Four elements of packing, rotated against each other by 90° with effective corrugation angle of 19°, were used. The model of packing of MELLAPAK 250Y made from polypropylene was employed. Simulations are carried out using CFX 10.0 to study the countercurrent two-phase flow [8]. Numerical result of the liquid flow shows the flow of liquid phase through the packing as well as the redistribution of liquid phase when liquid meets the joint of the packing elements. The flow behavior was also studied experimentally using X-ray tomographic visualization technique with contrast agent racer [9].

An exclusive investigation was performed to study the flow behavior of liquid film and rivulets on inclined planes. The volume-of-fluid-like model in CFX 5.0 was utilized to study the same. The influence of surface tension was taken into account, and the simulation studies were validated with experimental work [10–13].

A direct numerical simulation to study physical and reactive absorption in gas-liquid flow on structured packing was recently published. It also showed that the liquid side mass transfer is well predicted by the Higbie theory. The numerical results were compared to approximate solution presented in the literature [14, 15]. Van Baten and Krishna [16] studied the gas and liquid mass transfer in katapak-S structures using CFD simulations, and gas phase mass transfer was in good agreement with the theoretical correlation of Viva et al. [17], whereas the liquid phase mass transfer was onefold lesser than the correlation.

In the present work, the three-dimensional volume-of-fluid (VOF) model is presented to study the flow of liquid on the corrugated sheet of packing. The surface tension of the liquid has been taken into consideration using continuum surface (CSF) model. The transient simulations are performed using the geometric reconstruction scheme for interpolation near the interface, and SIMPLEC-based solver was utilized. The comparison between the simulation and experimental studies is also presented. Further, the study was extended to study the wetting behavior for two corrugated sheets of packing.

NUMERICAL DETAILS

Details of the Model

The simulations are carried out with the commercial tool ANSYS Fluent 12.0, ANSYS Inc [18].

Volume-of-Fluid (VOF) Model

The volume-of-fluid (VOF) model, which is one of the limiting cases of Euler-Euler homogenous model, is considered in this work. The VOF model considers that the gas and liquid phases are not interpenetrating. For each phase that is added, a variable is introduced. In each control volume, the volume fractions of all phases sum to unity.

The tracking of the interface(s) between the phases is accomplished by the continuity equation for the volume fraction of one (or more) of the phases. For the qth phase, this equation has the following form:

$$\frac{1}{\rho_Q} \left[\frac{\partial}{\partial t} \left(\alpha_q \rho_q \right) + \nabla \cdot \left(\alpha_q \rho_q \vec{v}_q \right) = S_{\alpha_q} \right.$$

$$\left. + \sum_{p=1}^{n} \left(\dot{m}_{pq} - \dot{m}_{qp} \right) \right],$$

$$(1)$$

where α_q is the volume fraction of the qth phase, ρ is the density, and S is the source term.

The volume fraction equation will not be solved for the primary phase; the primary-phase volume fraction will be computed based on the following constraint:

$$\sum_{q=1}^{n} \alpha_q = 1.$$

(2)

The volume fraction equation can be solved using explicit time discretization.

Explicit Discretization

In the explicit approach, finite difference interpolation schemes are applied to the volume fraction values that were computed at the previous time step:

$$\frac{\alpha_q^{n+1} \rho_q^{n+1} - \alpha_q^{n} \rho_q^{n}}{\Delta t} V + \sum_{f} \left(\rho_q U_f^n \alpha_{q,f}^n \right)$$

$$= \left[\sum_{p=1}^{n} \left(\dot{m}_{pq} - \dot{m}_{qp} \right) + S_{\alpha_q} \right] V,$$

(3)

where $(n+1)$ is the index of the new (current) time step, n is the previous time step, $\alpha_{q,f}$ is the face value of the qth fraction, V is the volume of the cell, and U_f is the volume flux through the face, based on normal velocity.

Interpolation near the Interface

Geometric reconstruction scheme was utilized in this work to interpolate variables near the interface between two phases.

In time-dependent VOF calculations, the time step used for the volume fraction calculation will not be the same as the time step used for the rest of the transport equations. ANSYS FLUENT will refine the time step for VOF automatically, based on the input for maximum Courant number allowed near the free surface. The Courant number (Co) is a dimensionless number that compares the time step in a calculation to the characteristic time of transit of a fluid element across a control volume:

$$Co = \frac{\Delta t}{\Delta x_{cell}/v_{fluid}},$$

(4)

where Δx_{cell} is the minimum cell dimension, v_{fluid} is the kinematic viscosity of the fluid, and Δt is the time step. The time step used in the simulation is $1*10^{-5}$ sec. In the region near the fluid interface, ANSYS FLUENT divides the volume of each cell by sum of the outgoing fluxes. The resulting time represents the time it would take the fluid to empty out of the cell.

Surface Tension

The VOF model can also include the effects of surface tension along the interface between each pair of phases. The influence of surface tension source term is taken into account by the continuum surface model (CSF) proposed by Brackbill [19]. The addition of surface tension to the VOF calculation results in a source term in the momentum equation.

All the simulations are performed under transient and laminar conditions. Detailed description of the equation and other assumptions has been already presented [20]. All the simulations are performed on IBM pSeries 690 supercomputers with SGI Altix XE 250 and in 32 parallel nodes of the HLRN (High Performance Computing Network of Northern Germany) at regional computing clusters available at Berlin.

Geometry and Dimensions

In order to study the flow behaviour in corrugated sheet of packing, three major geometries are considered in this work. Geometries are developed and meshed using ICEM CFD 12.0 [21]. List of geometries used in this work is presented in Table 1.

Table 1: Details of different geometries utilized in this work

S. no.	No. of sheets	Crimp apex	Perforations	Surface textures	Crimp angle
C1	One	Sinusoidal	No	Smooth	45°
C2	One	Sinusoidal	Yes	Smooth	45°
C3	Two	Sinusoidal	No	Smooth	45°

The detailed geometries with sinusoidal crimp are shown in Figure 3. Overall dimensions of the geometrical domain are 132 mm × 88 mm × 17 mm. The corrugation angle used in all the geometries is 45°, and the perforation is of 4 mm diameter with pitch of 10 mm along the length and width of the geometry, which is shown in Figure 3(d). The influence of the perforations was taken into consideration by changing the boundary conditions as described in Table 2. Figure 4 shows the detailed meshing of the geometry shown in Figure 3, and the mesh consists of 1143600 cells. Figure 5 shows the geometry in which two corrugated sheets are arranged as in practical applications, one corrugated sheet is turned 90° to the other.

Table 2: Boundary conditions used in the simulations

	Simulation without perforations	Simulation with perforations
Inlet	Velocity inlet	Velocity inlet
Outlet	Pressure outlet	Pressure outlet
Top	Pressure outlet	Pressure outlet
Bottom	Pressure outlet	Pressure outlet
Corruga-tion—base	Wall (CA)	Wall (CA)
Corruga-tion—holes	Wall (CA)	Interior
Sides	Symmetry	Symmetry

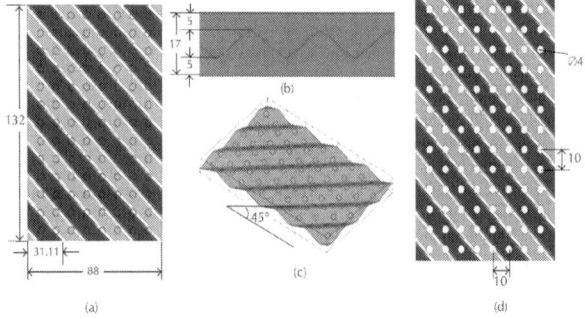

Figure 3: Schematic view of geometry of corrugated sheet of packing. (a) Top view: geometry without holes. (b) Side view. (c) Isometric view (d) Top view: geometry with holes. All dimensions are in mm.

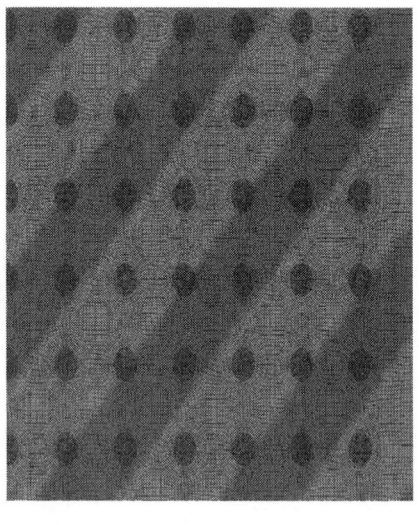

(a)

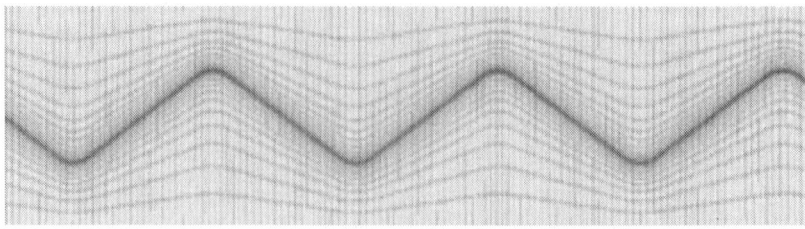

(b)

Figure 4: Meshes shown in detail for corrugated sheet of packing used in this work. (a) Top view. (b) Side view. (c) Isometric view.

(a)

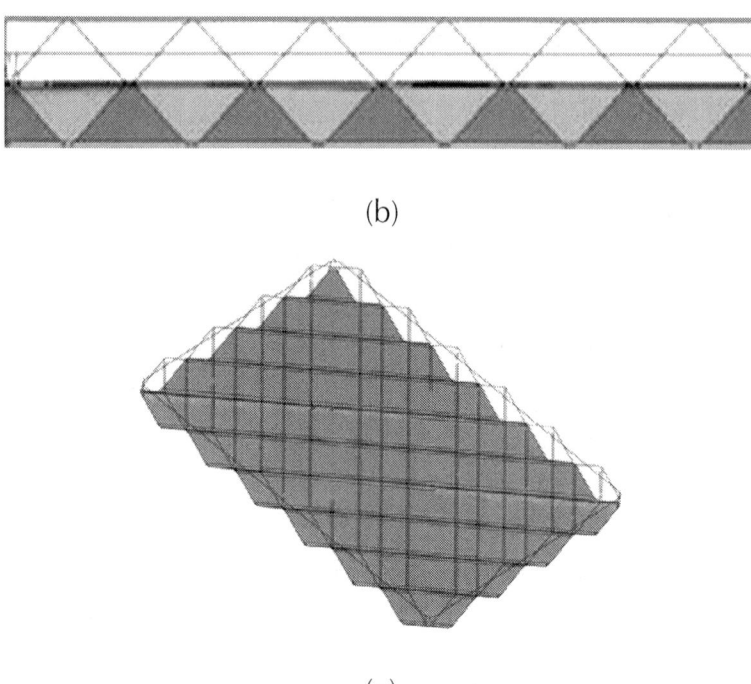

(b)

(c)

Figure 5: Schematic view of two corrugated sheets of packing with smooth crimp. (a) Top views (b) Side view. (c) Isometric view.

Boundary Conditions

Details of the boundary conditions used in the simulations are listed in Table 2. The influence of perforations has been brought in simulation by changing the boundary conditions of holes in the geometry. For the simulation with perforations, boundary conditions of the perforations will be assigned as interior, otherwise it will be assigned as wall. This method helps in using the same geometry for both the simulations and helps in comparing the influence later.

As described in Section 2.1, simulations are performed on three major geometries. For geometry C1 and C2, one inlet of 4 mm pipe was considered just to resemble real flow distributors used in industries. Three different flow rates are studied for each testing system. The details of the flow rates and Reynolds number presented in this work are shown in Table 3.

Table 3: Details of the flow rate and Reynolds number presented in this work

	Flow rate (mL/min.)	Re
Water	386	2033
	623	3291
Water-glycerol (45%)	386	486
	590	743
Silicon-oil (DC5)	241	230
	508	486

For simulation using geometry C3, 4 inlets each of 4 mm diameter were considered for study. The aim of using these 4 inlets is to study the maximum wetting possibility in the corrugated sheet of packing. Hence, very high flow rate of 811 mL/min was studied. Two major inlet positions as described in Section 3.3 are also considered.

Testing Systems

Three different fluids with difference in viscosity and contact angle were selected for this study. The details of the liquids are listed in Table 4. Water-glycerol (45 wt%) and water have similar contact angle but Water-glycerol has higher viscosity than water. Silicon-oil (DC5) has very low contact angle but similar viscosity to that of Water-glycerol (45 wt%). To capture the wetting, Rhodamine-B was used as coloring pigment in water and water-glycerol solution. As the color of the testing system is pink, the wetting can be studied without the help of UV-light. The usage of this coloring pigment has been studied earlier [22], and this will not influence any of the physical parameters of the testing system. For silicon-oil (DC5), Coumarin was used as a coloring agent, which gives blue reflections when studied with the help of UV-light. But the corrugated sheet must be coated with black color to capture the UV light.

Table 4: Properties of the testing system

	Viscosity η (mPa s)	Density ρ (kg/m³)	Surface tension σ (mN/m)	Contact angle θ (°)
Water	1	997	72.7	76.6
Water-glycerol (45 Wt%)	4.6	1113	70	69
Silicon-oil (DC5)	4.6	915	18.5	≈7

Experimental Setup

The flow diagram of the experimental setup is shown in Figure 6. Test liquids were pumped using pump (P01) from the solution tank (T01) to flow through the test cell (TC). Before flowing through the test cell, it passes through the flow indicator (FI) and buffer tank (B01). Buffer tank is used in order to avoid pulsations arising from the pump. A camera (C) is placed in the stand opposite to the corrugated sheet, which enables taking pictures.

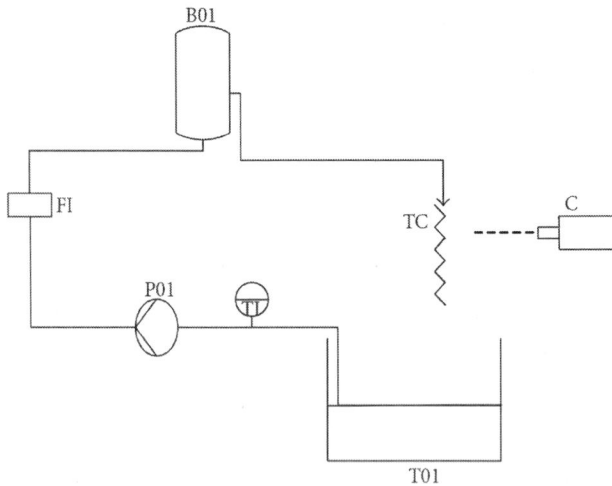

Figure 6: Experimental setup used in wetting studies. Sample picture from Water-Glycerol study. P01: pump; B01: buffer tank; TC: test cell; C: camera; FI: flow indicator; TI: thermocouple; T01: solution tank.

RESULTS AND DISCUSSION

Flow on Corrugated Sheet with Smooth Crimp Surface without Holes

In this section, comparison between experiment and simulation for flow of three different liquids on corrugated sheet of packing is presented. In Figure 7, a snapshot from simulation for a qualitative comparison of wetting behavior of silicon-oil (DC5), water, and water-glycerol on corrugated sheet is presented. It can be seen that the fluid with low contact angle, that is, silicon-oil (DC5), has more wetting in comparison to water and water-glycerol (45 wt%), which has very high contact angle. In other words, water, and water-glycerol system with high contact angle has a tendency to change its direction towards the corrugation than silicon-oil.

Silicon-oil

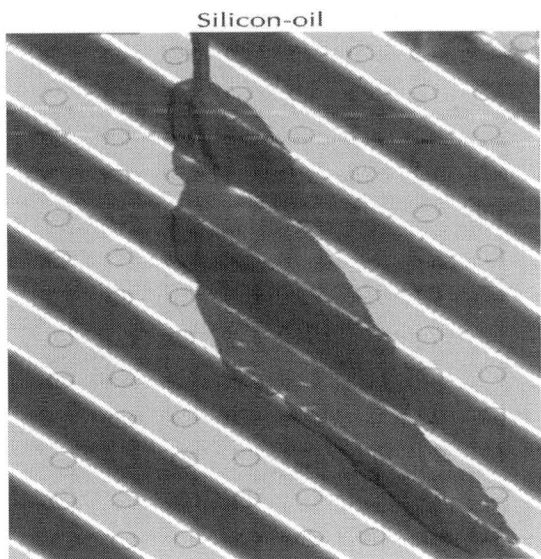

(a)

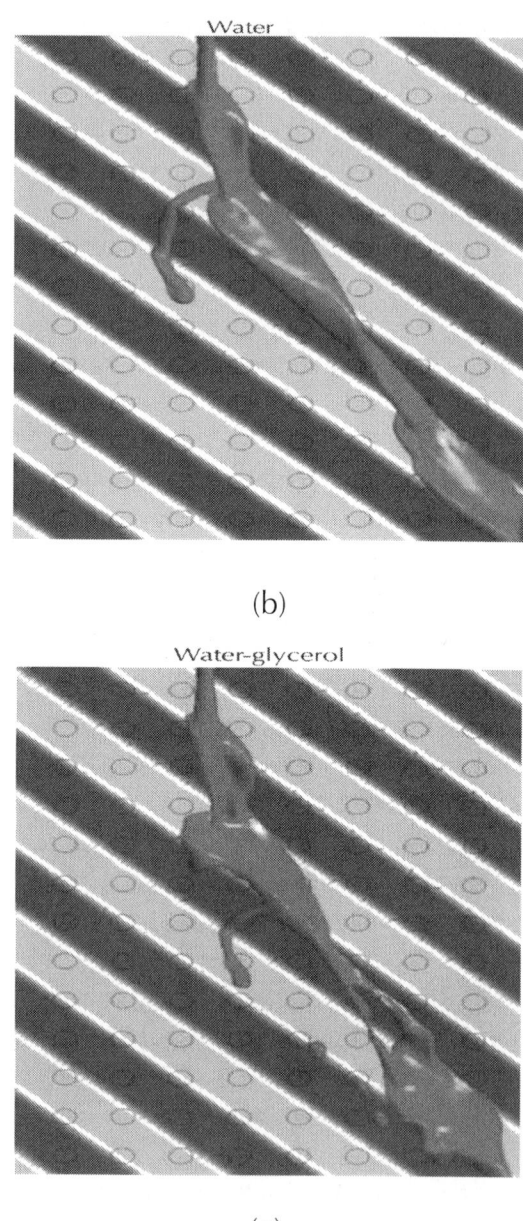

(b)

(c)

Figure 7: Snapshot from simulation for wetting behaviour of silicon-oil (DC5) and water and water-glycerol on corrugated sheet of packing without holes.

In Figures 8, 9, and 10, the flow and wetting behavior of water, water-glycerol, and silicon-oil (DC5) over corrugated sheet without perforations are shown in comparison with experimental studies. Geometry utilized in the simulation is similar to Figure 3. Liquid flows through circular inlets of 4 mm diameter, which reflects the distributors utilizing in real-time industrial applications, which stands at 90° from base. Experiments are performed on Montz B1-300 packing, which has textured surface. However, geometry used in the simulations is without microstructure on the surface. Hence, the differences are needed to be considered while comparing the simulation and experiments.

Simulation   Experiment

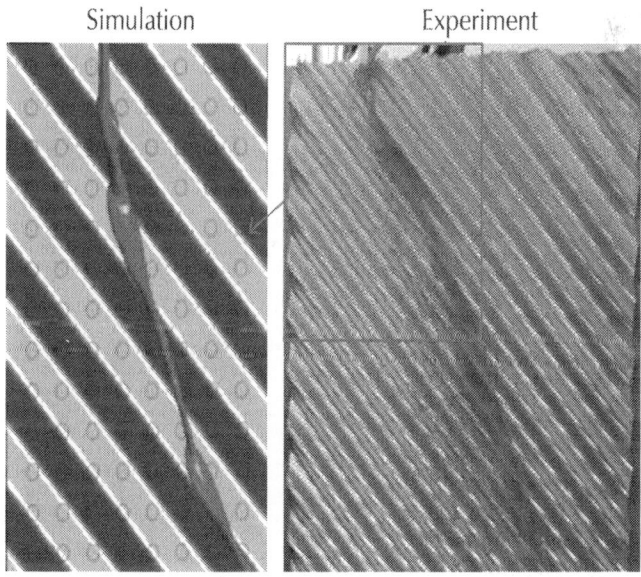

Figure 8: Comparison between simulation and experiment for flow of water on corrugated sheet of packing without holes. Flow rate: 386 mL/min. Rectangular box shown in red colour is the geometry considered in simulation studies.

Simulation   Experiment

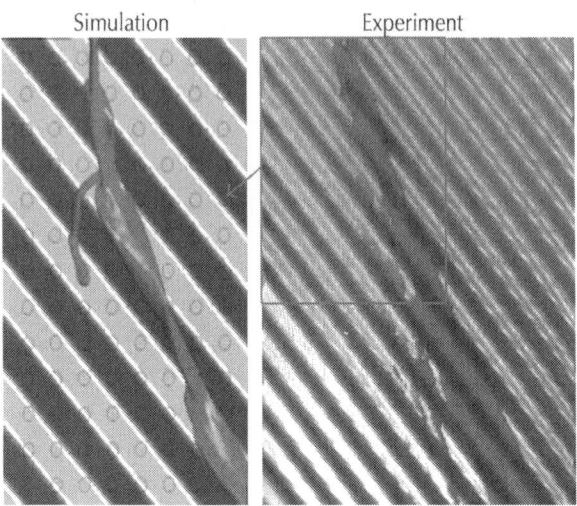

Figure 9: Comparison between simulation and experiment for flow of water-glycerol on corrugated sheet of packing without holes. Flow rate: 387 mL/min.

Simulation   Experiment

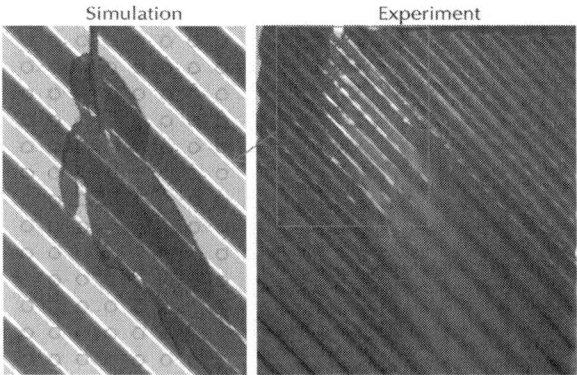

Figure 10: Comparison between simulation and experiment for flow of silicon-oil (DC5) on corrugated sheet of packing without holes. Flow rate: 241 mL/min.

As seen in Figures 8–11, the direction of the flow of liquid changes in the direction of corrugation, which helps to utilize the packing surface effectively and hence the wetting area increases.

When the liquid flows inside corrugation, the width of the liquid flow is more while it is lesser in the crimp of packing. In turn, the thickness of the liquid is more in the crimp of packing, which can be seen in Figures 8–11. This phenomenon is also in accordance with earlier experimental studies where the liquid hold-up is reported to be more around the crimp [23].

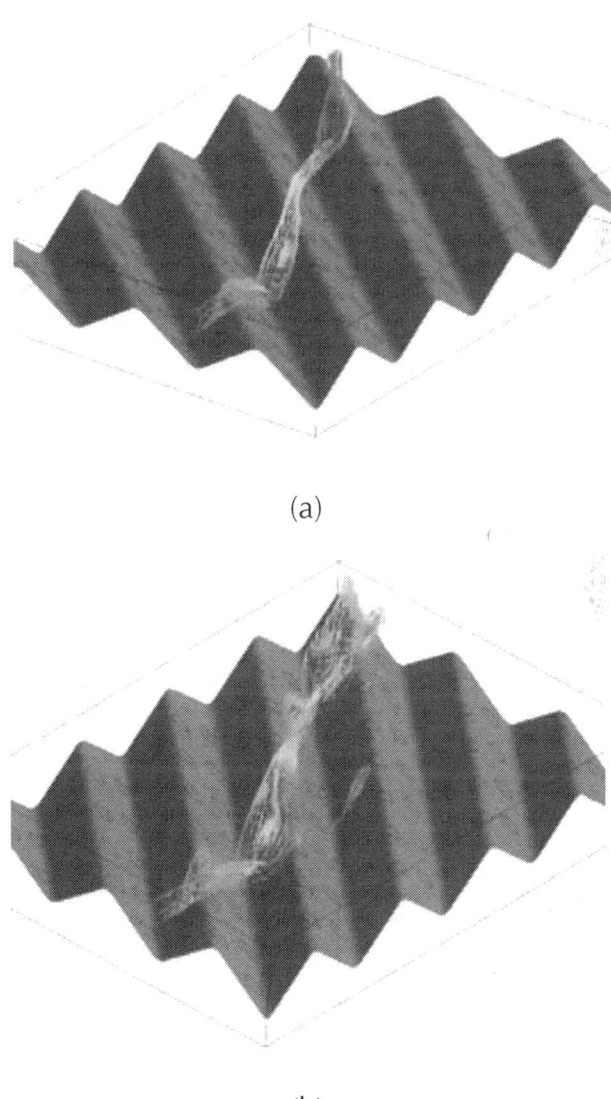

(a)

(b)

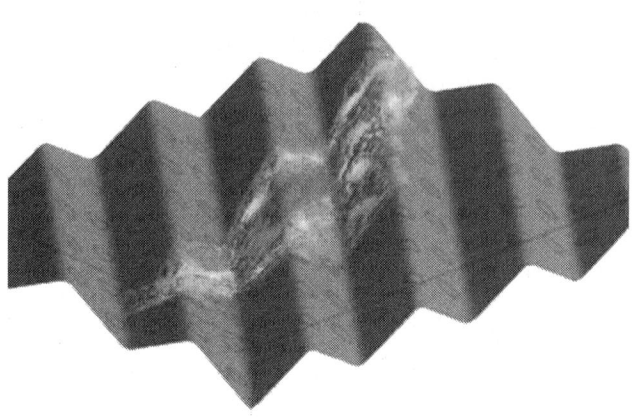

Figure 11: Velocity vectors for three liquids flowing through corrugated sheet of packing smooth crimp apex. (a) Water. (b) Water-glycerol. (c) Silicon-oil.

While comparing Figures 8 and 9 with Figure 10, the influence of contact angle can be noticed. The wetting of liquid with low contact angle, that is, silicon-oil (DC5), is higher than the wetting of water and water-glycerol Figures 9 and 10.

In Figure 11, closer view of the velocity vector is shown to understand the smooth flow of the liquid over the crimp of the corrugated sheet of packing. It is clear from this section that the liquid with low contact angle has very good wetting behaviour, which is also in accordance with our earlier studies with smooth inclined plate. The corrugation in the packing sheets helps the fluid to flow longer by changing the direction along the corrugation. This increases the contact time of the liquid-gas inside the packing.

Flow on Corrugated Sheet of Packing with Holes

In this section, the flows of different testing liquids on the corrugated sheet of packing with holes are shown. Figure 12–Figure 15 show the wetting on corrugated sheet of packing with smooth crimp surface and with holes for water, water-glycerol and silicon-oil. Figures 12 and 13 show the hydrodynamics of flow of water at two different flow rates. Figures 14 and 15 show the flow behavior of water-glycerol, and silicon-oil (DC5).

Simulation-front side

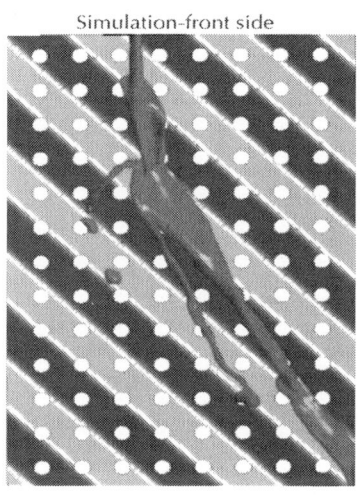

(a)

Experiment-front side

(b)

Simulation-rear side

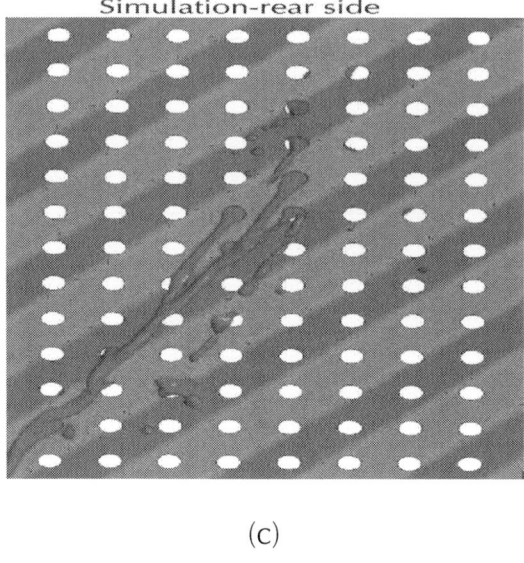

(c)

Figure 12: Comparison between simulation and experiment for flow of water on corrugated sheet of packing with perforations. Flow rate: 623 mL/min.

Simulation-front side

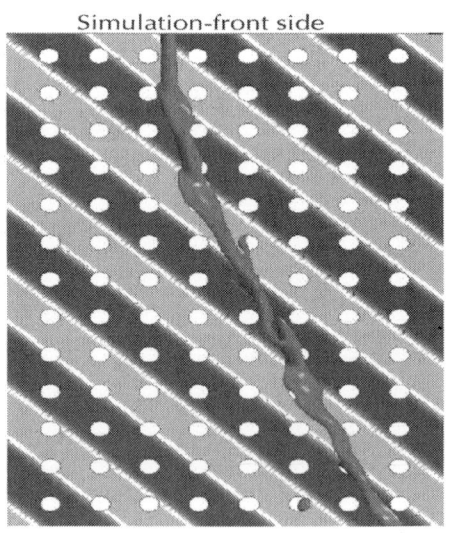

(a)

Experiment–front side

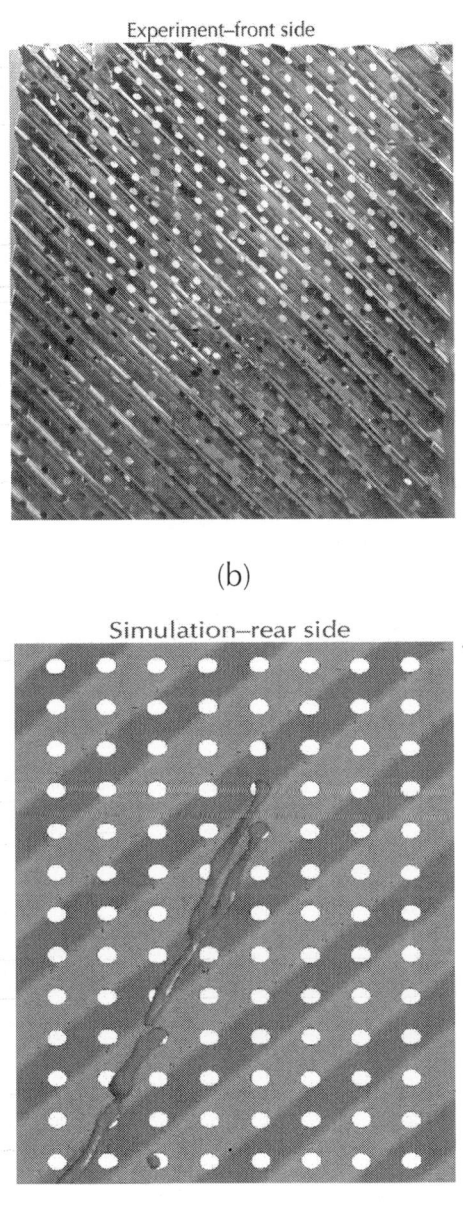

(b)

Simulation—rear side

(c)

Figure 13: Comparison between simulation and experiment for flow of water on corrugated sheet of packing with perforations. Flow rate: 386 mL/min.

Simulation–front side

(a)

Experiment

(b)

Simulation—rear side

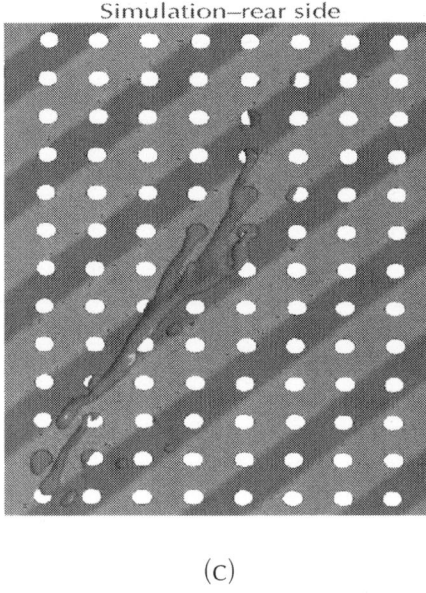

(c)

Figure 14: Comparison between simulation and experiment for flow of water-glycerol on corrugated sheet of packing with perforations. Flow rate: 590 mL/min.

Simulation—front side

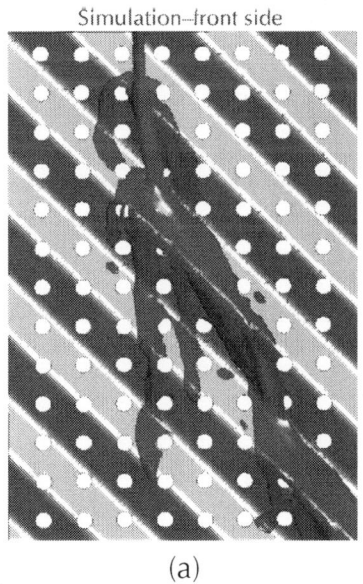

(a)

Experiment–front side

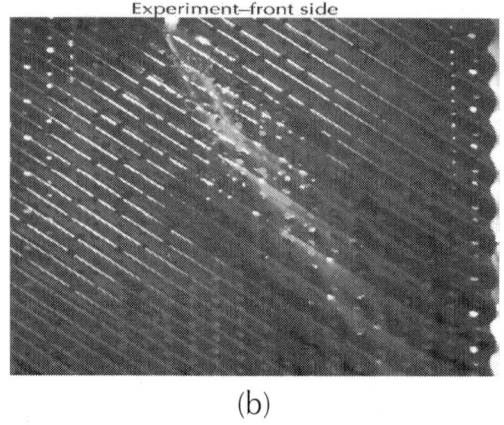

(b)

Simulation–rear side

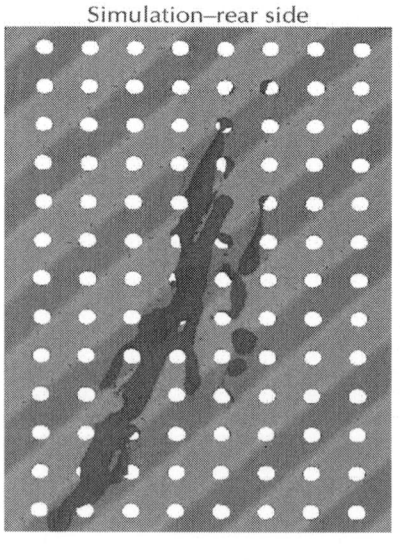

(c)

Figure 15: Comparison between simulation and experiment for flow of silicon-oil on corrugated sheet of packing with perforations. Flow rate: 508 mL/min.

The influence of perforations can be clearly seen while comparing Figure 8–Figure 10 with Figure 12–Figure15. Due to perforations, both sides of the corrugated sheet are wetted. However, the wetting area on the front side of the corrugated sheet considerably reduced in comparison to the sheet without perforations.

In Figure 12–Figure 15, comparison between experiment and simulation is also shown. The flow on the front side of the sheet is shown in comparison with simulation and experiment. It is very tedious to capture the flow on the rear side of the corrugated sheet experimentally. Hence, the flow on the rear of the packing is shown from CFD simulations.

By comparing Figures 12 and 13, the change in wetting due to change in flow rate can be observed. The wetting of water at high flow is more (Figure 12) compared to at low flow rate (Figure 13). On the other hand, it should be noticed that holes are utilized more in the low flow rate. Hence, the wetting on the rear side of the packing is more for low flow rate. So it looks like flow rate and holes play a contradictory role while wetting.

While comparing Figures 13 and 15, it can be understood that for high viscous fluids the flow wetting behavior has no big difference.

As seen in Section 3.1, silicon-oil has shown better wetting in the corrugated sheet of packing with perforations as well (Figure 16). The wetting on the front side of the packing is less in comparison with the packing without perforations. But the wetting on the rear side is considerably high. For silicon-oil, the change in flow rate did not make a huge difference in wetting the rear side of the sheet (not shown here).

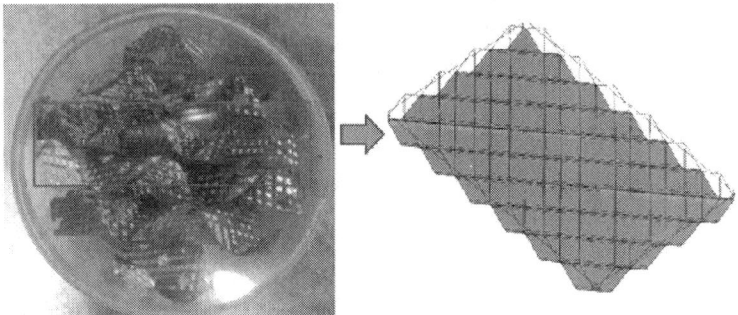

Figure 16: Comparison of domains used in simulation from real packing geometry.

Overall, the presence of perforations plays a key role in wetting and hence increases the contact area for the liquid gas interaction. Hence, in recent times, almost all the industrial packings have perforations. Now with this study, the wetting on the rear side of the packing can also be understood, which is usually cumbersome with experiments.

Two Corrugated Sheets of Packing

The major aim of this section is to study the extent of wetting on two corrugated sheets of packing. In reality, corrugated sheets of packing are arranged in such a way that one, sheet is placed 90° opposite to the other one, that is, corrugation lies in the opposite direction, helping the fluid to change its direction. Figure 16 shows the domain of the geometry used for simulation in comparison with real packing segment. As shown, only part of the packing segment is considered for simulation in order to understand the influence of the second sheet in the liquid hold-up and in the wetting pattern. The main region to be considered is the meeting points where two corrugated sheets touch each other, which are explained in Figure 17.

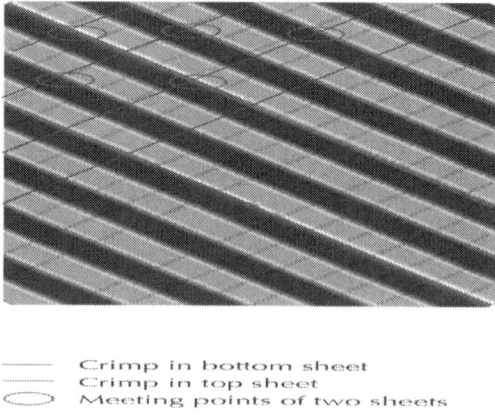

------- Crimp in bottom sheet
.... Crimp in top sheet
⊂⊃ Meeting points of two sheets

Figure 17: Explanation of meeting points from two corrugated sheets of packing and interest of our study.

Simulations are performed with the inlet conditions described as follows.

- Inlets are 4 mm diameter liquid distributors as seen in Section 3.1
- Four inlet distributors are considered to understand the maximum wetting possible with two corrugated sheets of packing.
- Two different inlet positions are considered as shown in Figure 18. Two inlet positions are chosen in such a way that one position is inside the corrugation of bottom sheet (Position 1) and the other position is outside the corrugation of bottom sheet (Position 2). By this, the influence of the meeting point due to the second corrugated sheet on the flow of liquid can be clearly seen.
- Silicon-oil with volumetric flow rate of 811 mL/min showed the maximum wetting in our earlier studies.

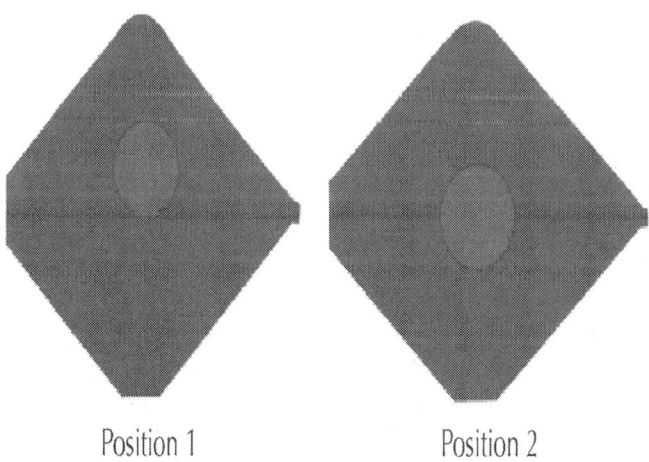

Position 1   Position 2

Figure 18: Two inlet positions used in the simulation.

Figures 19 and 20 show the isometric view of volume fraction of silicon-oil at different heights of the corrugated sheet of packing along the flow direction. It can be clearly seen that the liquid holds up near the criss-cross junction, that is, around the meeting point of the two sheets. Moreover, the hold-up is more near the inlet than in the outlet. This phenomenon of liquid holding up near this junction is seen in experimental study [23] performed using X-ray tomography for Mellapak 752.Y.

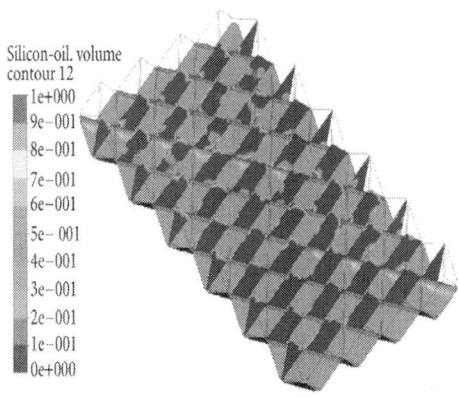

Figure 19: Contour of silicon-oil volume fraction at different heights of corrugated sheet of packing simulated along the flow direction for inlet position 1. Vol. Flow rate = 811 mL/min.

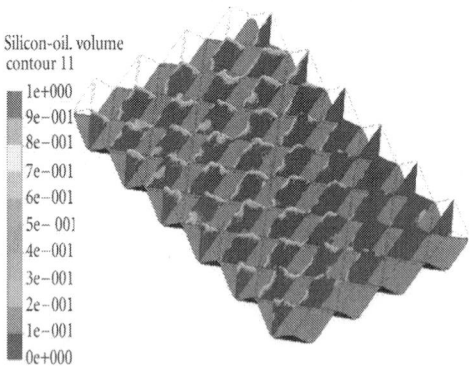

Figure 20: Contour of silicon-oil volume fraction at different heights of corrugated sheet of packing simulated in the flow direction for inlet position 2. Vol. Flow rate = 811 mL/min.

Figures 21 and 22 show the wetting of bottom and top sheets for simulation with two corrugated sheets of packing and for two different inlet positions mentioned earlier but for the same flow rate. It is very interesting to note that along with the liquid hold-up, small change in inlet positions makes a huge impact on wetting of the corrugated sheet. For position 1, where most of the portion of inlet lies inside the corrugation of top sheets wets only the top packing sheet and the bottom sheets remain without wetting. For position 2, where the inlet lies equally to both the corrugated sheets, wets more of the bottom sheet and considerably less of the top sheet compared to position 1.

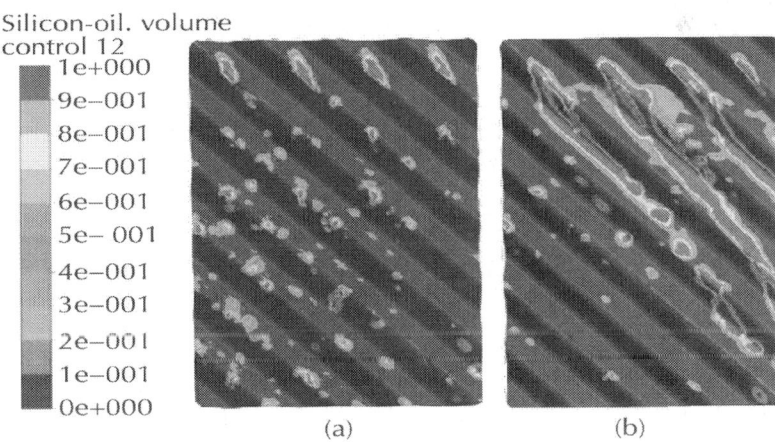

Silicon-oil. volume
control 12

1e+000	
9e−001	
8e−001	
7e−001	
6e−001	
5e− 001	
4e−001	
3e−001	
2e−001	
1e−001	
0e+000	

(a)   (b)

Figure 21: Wetted area on the bottom packing for two inlet positions. (a) Position 1. (b) position 2. Vol. Flow rate = 811 mL/min.

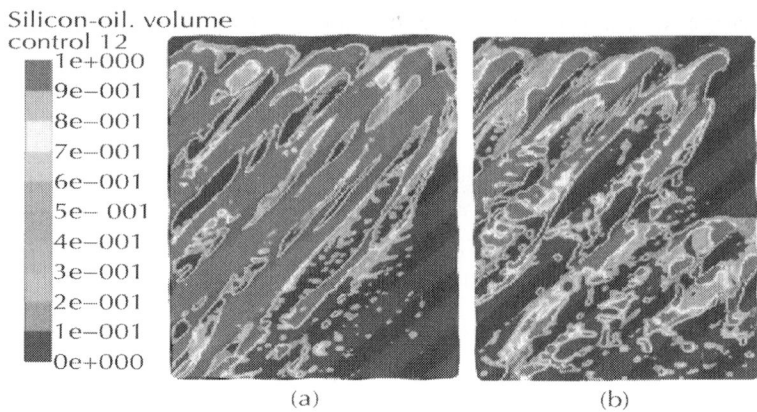

(a)   (b)

Figure 22: Wetted area on the top packing for two inlet positions. (a) Position 1. (b) Position 2. Vol. Flow rate = 811 mL/min.

In Table 5, the wetted areas of bottom and top corrugated sheets are shown. It is clear from position 1 that only 54% and 5.07% of top and bottom sheets are wetted, respectively. Moreover, in case of position 2, the wetted area is 15.36% and 39.06% for bottom and top sheets, respectively. Overall, only 55 to 60% of the packing area is utilized.

Table 5: Wetted area for bottom and top of the packing in % for silicon-oil (DC5)

	Position 1 (%)	Position 2 (%)
Bottom sheet	5.07	15.36
Top sheet	54.42	39.06
Total	59.49	54.42

As shown, only maximum of 60% of the packing area is utilised for wetting of testing system with low contact angle (which usually wets easily), for maximum inlet possible, that is, four inlets through four corrugations and relatively high flow rate of around 811 mL/min. It is very clear that around 40% of the area can be utilised and hence efficiency of the packing can be further improved. The influences of surface textures are not considered in the simulation, which can be considered further in simulation studies to understand the wetting behaviour better.

CONCLUSIONS

The three-dimensional VOF model is presented in this work in order to study the flow of liquid on the corrugated sheets of packing. Geometries with three different modifications in the element of the geometry have been studied. Three different testing fluids with different physical properties were considered. The simulations are performed in the 3D geometry resembling the corrugated sheet of packing. The simulations are compared with quantitative experimental studies. The corrugation changes the direction of the flow of fluid, and this increases the contact time of gas-liquid in the packing. As expected, the increase in flow rate also increased the wetting area. Testing mixture with low contact angle had better wetting.

The influence of perforations on wetting of the corrugated sheet was also shown. The wetting area on the front side of the sheet decreases while the liquid wets the rear side of the packing. Liquids flowing at low flow rate utilize the perforations more and wet both sides of the sheets. The increase in flow rate counter affects the presence of perforations. Due to high flow rate, perforations are not utilized completely hence the wetting on the rear side of packing was lesser compared to low flow rate. The second corrugated sheet was introduced to simulation domain to understand the influence of criss-cross junctions where both the sheets touch each other. Liquid hold-up near the junctions of two sheets was high, as noticed in some experimental studies performed using X-ray tomography [17]. It is also shown that the minor change in position of the inlet distributors considerably changes the wetting area and the flow direction of the liquid. Even for the simulation performed with the maximum number of inlets and for liquid with low contact angle, the complete area of packing was not utilized. This clearly proves that the efficiency of the existing packing can be further increased. The wetting of the corrugated sheet of packing can be further increased by textures on the surface. This theoretical work can be extended to study the wetting behavior on the corrugated sheet with microstructures on the surface of the packing. Various surface textures are possible as explained earlier in Section 1.2. The influence of the different 2D and 3D microstructures can be studied. Considering the validity of this model, dimension (nm to mm) of the microstructure can be optimized to achieve maximum wetting and hence high mass transfer. It is highly recommended to study the influence of microstructures using the CFD simulations, which will help to develop new surface textures to utilize the surface of packing completely. With the validated model, experiment efforts can be reduced but cannot be avoided completely. This study can also be further extended to study the transport processes in distillation and absorption.

ACKNOWLEDGMENTS

The authors thank the management committee of Evonik Stiftung for funding this project and are grateful to HLRN Berlin for using parallel computing in clusters. They also thank Mr. Martin Kohrt for his technical assistance in experimental study and Dr. Ilja Ausner for supplying the packing sheets for experiments.

REFERENCES

1.   H. Kister, Distillation Design, McGraw-Hill, 1992.
2.   S. J. Luo, W. Y. Fei, X. Y. Song, and H. Z. Li, "Effect of channel opening angle on the performance of structured packings," Chemical Engineering Journal, vol. 144, no. 2, pp. 227–234, 2008.
3.   Ž. Oluji , A. F. Seibert, and J. R. Fair, "Influence of corrugation geometry on the performance of structured packings: an experimental study," Chemical Engineering and Processing, vol. 39, no. 4, pp. 335–342, 2000.

4. J. T. Davies and K. V. Warner, "The effect of large-scale roughness in promoting gas absorption,"Chemical Engineering Science, vol. 24, no. 2, pp. 231–240, 1969.

5. L. Zhao and R. L. Cerro, "Experimental characterization of viscous film flows over complex surfaces,"International Journal of Multiphase Flow, vol. 18, no. 4, pp. 495–516, 1992. ·

6. S. Shetty and R. L. Cerro, "Flow of a thin fiom over a periodic surface," International Journal of Multiphase Flow, vol. 19, no. 6, pp. 1013–1027, 1993.

7. S. Shetty and R. L. Cerro, "Spreading of a liquid point source over a complex surface," Industrial and Engineering Chemistry Research, vol. 37, no. 2, pp. 626–635, 1998.

8. B. Mahr, "Numerisches Berechnen und tomographisches Messen zwei-phasiger Strömugsfelder in geordneten Schichtungen," in Institut für Verfahrenstechnik, Leibnitz Universität Hannover, Hanover, Germany, 2007.

9. B. Mahr and D. Mewes, "X-ray tomographic visualization of liquid spreading in structured packings using contrast-agent tracer," in Proceedings of the 13th International I leat Transfer Conference (IHTC ‹06), Sydney, Australia, 2006.

10. J. U. Repke, I. Ausner, S. Paschke, A. Hoffmann, and G. Wozny, "On the track to understanding three phases in one tower," Chemical Engineering Research and Design, vol. 85, no. 1, pp. 50–58, 2007.

11. A. Hoffmann, I. Ausner, J. U. Repke, and G. Wozny, "Fluid dynamics in multiphase distillation processes in packed towers," Computers and Chemical Engineering, vol. 29, no. 6, pp. 1433–1437, 2005.

12. A. Hoffmann, I. Ausner, J. U. Repke, and G. Wozny, "Detailed investigation of multiphase (gas-liquid and gas-liquid-liquid) flow behaviour on inclined plates," Chemical Engineering Research and Design, vol. 84, no. 2, pp. 147–154, 2006.

13. Y. Xu, S. Paschke, J. U. Repke, J. Yuan, and G. Wozny, "Portraying the countercurrent flow on packings by three-

dimensional computational fluid dynamics simulations," Chemical Engineering and Technology, vol. 31, no. 10, pp. 1445–1452, 2008.

14. Y. Haroun, D. Legendre, and L. Raynal, "Direct numerical simulation of reactive absorption in gas-liquid flow on structured packing using interface capturing method," Chemical Engineering Science, vol. 65, no. 1, pp. 351–356, 2010.

15. Y. Haroun, D. Legendre, and L. Raynal, "Volume of fluid method for interfacial reactive mass transfer: application to stable liquid film," Chemical Engineering Science, vol. 65, no. 10, pp. 2896–2909, 2010.

16. J. M. van Baten and R. Krishna, "Gas and liquid phase mass transfer within KATAPAK-S structures studied using CFD simulations," Chemical Engineering Science, vol. 57, no. 9, pp. 1531–1536, 2002.

17. A. Viva, S. Aferka, D. Toye, P. Marchot, M. Crine, and E. Brunazzi, "Determination of liquid hold-up and flow distribution inside modular catalytic structured packings," Chemical Engineering Research and Design, vol. 89, no. 8, pp. 1414–1426, 2011.

18. Ansys Fluent 12.0 User Guide, 2009.

19. J. U. Brackbill, "A continuum method for modeling surface tension," Journal of Computational Physics, vol. 100, no. 2, pp. 335–354, 1992.

20. K. Subramanian, S. Paschke, J. U. Repke, and G. Wozny, "Drag force modelling in CFD simulation to gain insight of packed columns," in Proceedings of the 9th International Conference on Chemical and Process Engineering, (ICheaP ‹09), pp. 561–566, May 2009. ·

21. Ansys ICEM CFD Manual 12.1, 2009.

22. S. Paschke, Experimentelle Analyse ein- und zweiphasiger Filmströmungen auf glatten und strukturierten Oberflächen, TU Berlin, Berlin, Germany, 2011.

23.   A. Viva, S. Aferka, E. Brunazzi, P. Marchot, M. Crine, and D. Toye, "Processing of X-ray tomographic images: a procedure adapted for the analysis of phase distribution in MellapakPlus 752.Y and Katapak-SP packings," Flow Measurement and Instrumentation, vol. 22, no. 4, pp. 279–290, 2011

Citations

CHAPTER 1

Deify Law, Francine Battaglia, Theodore J. Heindel, Model validation for low and high superficial gas velocity bubble column flows, Chemical Engineering Science, Volume 63, Issue 18, September 2008, Pages 4605-4616, ISSN 0009-2509, http://dx.doi.org/10.1016/j.ces.2008.07.001.

CHAPTER 2

Carl Montgomery (2013). Fracturing Fluids, Effective and Sustainable Hydraulic Fracturing, Dr. Rob Jeffrey (Ed.), ISBN: 978-953-51-1137-5, InTech, DOI: 10.5772/56192.

CHAPTER 3

Jinli Zhang, Shuangqing Xu, Wei Li, High shear mixers: A review of typical applications and studies on power draw, flow pattern, energy dissipation and transfer properties, Chemical Engineering and Processing: Process Intensification, Volumes 57–58, July–August 2012, Pages 25-41, ISSN 0255-2701, http://dx.doi.org/10.1016/j.cep.2012.04.004.

CHAPTER 4

A. Abubakar, T. Al-Wahaibi, Y. Al-Wahaibi, A.R. Al-Hashmi, A. Al-Ajmi, Roles of drag reducing polymers in single- and multi-phase flows, Chemical Engineering Research and Design, Volume 92, Issue 11, November 2014, Pages 2153-2181, ISSN 0263-8762, http://dx.doi.org/10.1016/j.cherd.2014.02.031.

CHAPTER 5

Kumar Subramanian and Günter Wozny, "Analysis of Hydrodynamics of Fluid Flow on Corrugated Sheets of Packing," International Journal of Chemical Engineering, vol. 2012, Article ID 838965, 13 pages, 2012. doi:10.1155/2012/838965.

Index